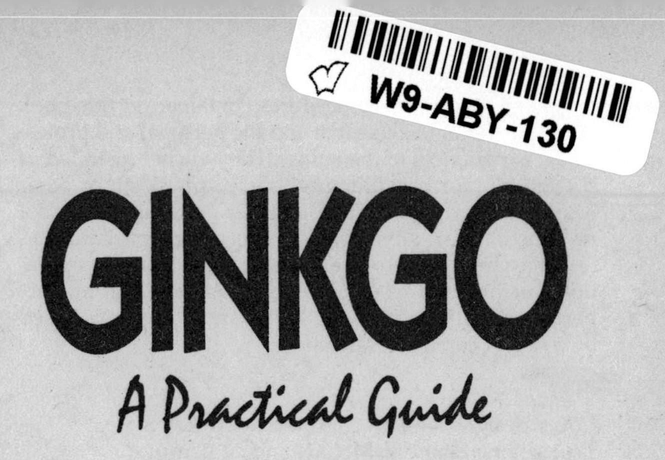

GINKGO

A Practical Guide

GEORGES HALPERN, MD, PhD

AVERY PUBLISHING GROUP
Garden City Park • New York

The information and procedures contained in this book are based upon the research and the personal and professional experiences of the author. They are not intended as a substitute for consulting with your physician or other health care provider. The publisher and author are not responsible for any adverse effects or consequences resulting from the use of any of the suggestions, preparations, or procedures discussed in this book. All matters pertaining to your physical health should be supervised by a health care professional.

Cover design: William Gonzalez and Rudy Shur
Typesetters: Elaine V. McCaw and Al Berotti
In-house editor: Eric Kraft
Page 14 photo credit: © 1996 Steven Foster

Avery Publishing Group, Inc.
120 Old Broadway
Garden City Park, NY 11040
1-800-548-5757

ISBN: 0-89529-900-3

Printed in the United States of America

10 9 8 7 6 5 4 3 2 1

Contents

Acknowledgments

Thanks to Jean-Michel Mencia-Huerta, Philippe M. Guinot, Pierre Braquet, and many more who convinced me of the benefits of ginkgo.

Also thanks to my wife Emiko, who cares for a wonderful ginkgo tree in Portola Valley, California, where our family enjoys it.

Introduction to Ginkgo

Ginkgo biloba products are sold in stores throughout the United States and Canada, but most Americans think of ginkgo as nothing more than a beautiful ornamental tree. Have we been missing out on important knowledge that could make our lives happier and healthier? I believe we have. That belief made me decide that it was time for Americans to discover the health benefits of *Ginkgo biloba* extract. So I decided to write this book to make the facts and history of this great herb known.

JAMES RESTON'S PERSONAL DISCOVERY OF CHINESE MEDICINE

The modern Western world discovered the remarkable potential of Chinese medicine in the early 1970s when, after a quarter-century of isolation, Chinese officials invited tourists to visit this mysterious, fabled land. Visitors from Europe and America were thrilled by the immensity of the Great Wall, the charm of China's gilded pagodas, and the ornate palaces of nearly-mythical emperors from vanished dynasties, but Westerners made no greater discovery than Chinese medicine.

One of the most famous visitors to the newly opened China was James Reston, a vice president of *The New York Times* and a columnist for that newspaper. Reston was stricken with acute appendicitis while visiting Beijing in 1971. He required an appendectomy to save his life. Doctors at the Anti-Imperialist Hospital removed his appendix after injecting him with lidocaine and benzocaine, which anesthetized his stomach area.

The second night after the surgery, Reston experienced great discomfort. With his approval, Chinese doctors rejected the idea of sedating him with drugs. Instead, they dulled his nerves with acupuncture, a therapeutic process in which specific points of the body are stimulated to promote healing. The doctors inserted three long, thin needles in his right elbow and below his knees.

Reston reported that insertion of the needles "sent ripples of pain through my limbs, which at least had the effect of diverting my attention from the distress in my stomach." One of Reston's doctors burned two pieces of an herb called *ai*, which to Reston looked like the stub of a used cigar, while twirling the needles. The writer said, "I remember thinking that it was a rather complicated way to get rid of gas on the stomach, but there was a noticeable relaxation of the pressure and distension within an hour and no recurrence of the problem thereafter." Reston, who was mentally alert throughout the operation, and the later acupuncture procedure, wrote a story about the experience that appeared on page one of *The New York Times* on July 26, 1971.

Readers thrilled to the story, which seemed to combine ancient wisdom with the magic arts. Alien doctors from Mars could not have inspired more interest than the Chinese physicians who treated Reston in Beijing with a method that had been known to their predecessors for thousands of years. However, as dramatic as acupuncture is, it is just one aspect of the ancient Chinese medical system, of which herbal treatments are an equally important branch.

THE TRADITIONAL USE OF PLANTS IN MEDICINE

All civilizations have traditions of using plants to promote healing. Indeed, most of these plants were the basis of the development of modern drugs. Aspirin, which is a chemical copy of the active analgesic chemicals in willow bark, is the most common example. Even today, about one-quarter of the drugs prescribed in the United States contain at least one active ingredient that is based on plant material.

Plants used in Chinese medicine are the basis for many common drugs in the West. Ephedrine, which is extracted from the herb *ma huang*, is used to ease the difficult breathing of asthma sufferers. Digitalis, a drug used to enable the heart to pump more blood—in other words, to increase cardiac output—is an extract of *mao ti huang*, which English-speaking people call foxglove. The history of digitalis shows how modern Western doctors adjust the ancient secrets of herbal medicine to suit their own biomedical systems. For centuries, herbalists used purple foxglove to treat dropsy (now called edema), which is a swelling due to the accumulation of fluid in the tissues, a result of a sluggish heart rate. In 1785, chemists found that the active ingredient of this plant was in the leaves of the foxglove. They were able to distill this and use it to create digitalis, which helped in the treatment of other heart ailments as well. Two related drugs derived from this plant, digoxin and digitoxin, are used to treat congestive heart failure around the world.

Now another herb, which has the potential to help in the treatment of an even wider range of ailments, is becoming known in the West. It is *Ginkgo biloba*.

Changing American Attitudes Toward Alternative Medicine

More than a quarter of a century after James Reston had his astonishing experience, Americans in general, and the American medical establishment in particular, have come to recognize the value of alternative medicine. In fact, a recent study showed that people who turn to alternative medical systems and practices tend to be better educated and more prosperous than those who do not. In 1991, the National Institutes of Health established an Office of Alternative Medicine (OAM), with a budget of $2 million, to "more adequately explore unconventional medical practices." (This office is now called the Office of Complementary and Alternative Medicine.) As the United States Senate Appropriations Committee noted when authorizing funding for the OAM, "Many routine and effective medical procedures now considered commonplace were once considered unconventional and counter-indicated. Cancer radiation therapy is such a procedure that is now commonplace but once was considered to be quackery."

The OAM sponsored a series of workshops on a variety of alternative medical practices, from acupuncture to herbs, as a first step in evaluating their value for the public. The authors of the report that arose from these workshops, *Alternative Medicine: Expanding Medical Horizons,* noted in its preface: "While the dominant system of health care in the United States—often called 'conventional medicine,' or biomedicine—is extremely effective for treating infectious diseases and traumatic injuries, it is often ill equipped to handle complex, multifaceted chronic conditions. One reason is that over the years, conventional medicine has increasingly emphasized finding a single 'magic bullet' solution for each condition or disease it confronts. The reality is that many chronic conditions are not amenable to such one-dimensional solutions."

The report went on say, "Such complex conditions require equally multifaceted treatment approaches," and stated that it was much cheaper "to prevent these medical problems than to treat the symptoms and consequences with surgery and expensive drugs."

Researchers reviewed the use of *Ginkgo biloba* in Europe. Noting that Europeans used it to combat the aging process, the OAM report stated that *Ginkgo biloba* extract has been shown to increase blood flow to the brain and to improve the brain's ability to cope with oxygen deficiency. *Alternative Medicine: Expanding Medical Horizons* stated, "No other known circulatory stimulant, natural or synthetic, has selectively increased blood flow to disease-damaged brain areas."

Alternative health care is now being recognized in America to the extent that some of the most popular health maintenance networks will reimburse patients who see acupuncturists and chiropractors as readily as they reimburse those who see general practitioners.

Despite all this, relatively little information on Chinese herbal medicine has reached the West. In fact, the American public probably knows less about the way Chinese doctors use plants than about acupuncture. This lack of awareness is due in part to language difficulties, but the greater obstacle has been a fundamental difference between the way Asian teachers and Western students approach the study of medicine. The Westerner believes that he or she must know the entire theory of the disease and the intellectual basis of the treatment before he or she can treat an individual patient successfully. Physicians in Asia believe in the practical application of basic principles. The Western medical student works diligently in the lab and the library for years before touching a patient. The Asian apprentice begins the first day of study beside his or her teacher, treating patients and learning directly from their symptoms and complaints.

Now, through books like this one, the West is becoming increasingly able to learn about these seemingly exotic treatments in a way that is relatively straightforward and easy to understand.

Some Reasons for Choosing Herbal Medicines

With our newspapers and airwaves filled with news of new scientific breakthroughs every day, why return to plant medicine? One reason is that plants offer a gentle, cumulative effect that Western science has bypassed in its search for quick eradication of disease or suppression of symptoms. Western drugs act quickly and are helpful when an illness is acute and severe. Herbs release their ingredients slowly into the system and are usually mild. They are particularly suitable for chronic problems where a slow approach is preferable. A second, equally important reason is that herbs offer an inexpensive way to treat large numbers of people. In the West, particularly in the United States, people who need medical care most—children, the elderly, the chronically ill—are least able to pay for treatment themselves. Chinese herbal medicine has given the Western world new and important "raw materials," such as ginkgo, with which to treat disease. Western scientists take these raw materials to their labs, prepare standardized extracts from the leaves, and turn them into tablets, liquids, and intravenous preparations.

A BRIEF HISTORY OF GINKGO BILOBA

The ginkgo is the oldest living tree species. It was prevalent during the Triassic period of the Mesozoic era, when dinosaurs roamed the earth, and some scientists think of it as a living fossil. Its resistance to parasites, insects, and pollution ensure it an astonishing natural life span of as long as a thousand years. It was common in many parts of the globe during the age of dinosaurs, but it died out in most places during the Ice Age, surviving only in Asia. In China, it was cultivated as a sacred tree within temple walls.

Traditional Chinese Uses for Ginkgo

For thousands of years, the Chinese have used ginkgo seeds as a treatment for conditions that traditional medicine considered "wet and runny," such as asthma and tuberculosis. They also used them as an aid to digestion and to expel intestinal worms. They created a tincture or extract as an aid to circulation and mental performance. The coatings of the seeds were used as an insecticide. As early as 2800 B.C.E., they used ginkgo leaves to restore memory and ease breathing problems.

European Acceptance of Ginkgo in Medicine

While the Chinese revere ginkgo, Europeans have come to rely upon it. *Ginkgo biloba* extract is among the most commonly prescribed drugs in France and Germany, where it is used to fight many of the common symptoms of aging. *Ginkgo biloba* is registered as a drug in those countries for the treatment of organic brain disorders. It accounts for 1 percent of all prescription medicines in Germany and 4 percent in France, and is sold in lower dosages over the counter. In Germany, ginkgo is also licensed for the treatment of "cerebral insufficiency." This term refers to a host of problems ranging from impaired memory, dizziness, and ringing in the ears to headaches, nervousness, and anxiety. The extract has also been approved as a supplemental treatment for certain kinds of hearing loss and for leg cramps and numbness due to poor circulation.

An observer of European interest in *Ginkgo biloba,* David B. Mowrey, Ph.D., director of the American Phytotherapy Research Lab in Salt Lake City, Utah, wrote in *The Scientific Validation of Herbal Medicine* (Keats Publishing, 1990), "The research on ginkgo and Alzheimer's is producing extremely good results in France and Germany. It seems that the earlier you catch the disease, the greater the chance that you can reverse it by taking ginkgo extract."

A Summary of Ginkgo's Benefits

The key to ginkgo's benefits is its ability to improve circulation to every area of the body, including the brain. Today we know that *Ginkgo biloba* helps to keep the blood vessels supple and elastic, and thus helps prevent heart and circulatory problems. Research indicates that it is effective in the early stages of Alzheimer's disease and strokes. By improving blood flow, ginkgo enhances the body's ability to bring nourishment to every part of the body. The improved blood flow that ginkgo fosters also helps heal disorders with a circulatory basis, ranging from ringing in the ears to numbness in the toes.

Ginkgo also fights free radicals. Free radicals are highly reactive atoms and molecules. When they are absorbed into the body—whether by breathing in toxins, consuming harmful foods such as fats, or through other means—free radicals can damage the body and contribute to many diseases. Some free radicals can react with cellular structures, causing immediate or latent damage. Free radicals are widely believed to play a role in degenerative diseases such as cancer, in the aging process, and in the development of Alzheimer's disease. Antioxidants such as *Ginkgo biloba* are substances that "scavenge" free-radicals, reacting with them to leave harmless molecules in their place.

A Brief Survey of Research Supporting the Benefits of Gingko

As *Alternative Medicine: Expanding Medical Horizons,* the OAM report to the National Institutes of Health, demonstrated, Western medicine is catching on to the benefits of herbs and adopting them in alternative treatments. *Ginkgo biloba* extract is among the most well studied plant-based medicines in Europe and the United States. It can protect against various disorders and often ease their symptoms. Its antioxidant properties fight free radicals and, I believe, are the basis of ginkgo's multiple beneficial effects, which include improved circulation and improved health of the brain, the heart, and the retina of the eye. I will discuss these and other advantages of ginkgo in subsequent chapters and provide supporting research. Here I will consider a few outstanding studies.

In one breakthrough study, German scientists in 1992 administered high dosages of *Ginkgo biloba* extract (240 milligrams daily) to twenty patients with various diseases that led to dangerous blood clotting. These patients, who were not hospitalized, had coronary heart disease, hypertension, high cholesterol, and diabetes. After twelve weeks of treatment, clotting factors—that is, the substances in blood that promote clotting—had decreased in every one of the patients taking *Ginkgo biloba,* leading the researchers to report in the medical journal *Fortschrift Medizin* that "the medication can thus positively influence these cardiovascular risk factors over the long term."

Possibly the most famous study of the effects of *Ginkgo biloba* extract in the treatment of cerebral disorders brought on by the aging process was reported in the September 25, 1986, issue of the French medical journal *La Presse Médicale*. French researchers developed a specific scale of seventeen items to evaluate 166 "geriatric" patients in several centers. These seventeen markers included vivacity, short-term memory, disturbances in orientation, anxiety, depression, emotional stability, initiative, cooperation, sociability, personal care, ability to walk, appetite, vertigo, fatigue, headache, sleep, and ringing in the ears. Subjects improved in all cases after taking *Ginkgo biloba* extract for three months, and they continued to improve over time.

In a German study published early in 1996, ginkgo proved to be beneficial in the early stages of Alzheimer's disease. The study involved 216 patients suffering from mild to moderate symptoms of Alzheimer's. The patients were divided into two groups. For a month, the patients in one group were treated with 240 milligrams of *Ginkgo biloba* extract daily, and the patients in the other were given a daily placebo. At the end of the period, the subjects were tested for mental, behavioral, and motor skills. Those who had taken ginkgo showed a great increase in mental alertness and an improvement in mood, but the patients in the placebo group showed little improvement.

In many cases, *Ginkgo biloba* has not proven more effective than drugs in the treatment of similar conditions. However, because ginkgo has no serious side effects and no problems have been observed arising from interactions with other medications that a patient may be taking, this herbal product may be preferable to strong drugs in those cases.

Ginkgo's Role in Alleviating Problems of Aging

Before the improvement of hygiene and control of infectious

disease, people on the whole died at younger ages than they do today. When I was a boy, I did not know many people over the age of sixty. Now, because of all the medical progress we have made, we have a growing population over the age of sixty-five. The average age in the United States is on the rise, with people over sixty-five expected to account for slightly more than 12 percent of the population by the year 2000. Now the problem is not death, but aging. It is a growing problem because our society is at the same time aging and less tolerant of aging. Our older people do not have a role. In contrast, in Japan, it is still common to have four generations living together. The great-grandmother may clean the house and take care of the baby. She has a purpose, and if her memory begins to fail, it does not matter so much because her family can help her remember.

The fast pace of our society does not make accommodations for people who slow down. It is true that up to a certain point older people do better today than people their age did a generation or two ago, but after a certain age people seem to go into a quick decline, a sudden shocking drop rather than a slow decline with which others can cope. In fact, this decline is a slow process that goes largely unnoticed until it passes a point where it cannot be ignored. It's like a toothache that begins as a tolerable discomfort and slowly increases until it "suddenly" becomes intolerable.

In our culture today, either you are seen as normal or you are seen as not normal, and troubles associated with aging can make an older person seem to be not normal. "Troubles associated with aging" is a big black box into which we put a lot of symptoms, including the decline of mental functions. Modern life makes people intolerant of those who are not performing well intellectually, whether it is due to aging or some other problem. We need to prevent the degradation of our mental functions and to treat any such problems once they begin to occur. We need to do this early enough so that we ourselves or those we love do not need to be institutionalized or treated with expensive drugs. Because the best environment is the family, we must help people to be able to function well enough so that they can stay in the family, be part of it and contribute to it. That is impossible if they are demented. The scientific evidence suggests that *Ginkgo biloba* can postpone the day, or possibly even prevent the day, when mental function is seriously impaired.

Ginkgo has been tested in humans, not just in animals and in test tubes. In scientific studies comparing *Ginkgo biloba* to seventeen drugs used in the treatment of vascular (blood vessel-related) disorders, ginkgo was the only one that was found to be effective against all of the following problems associated with aging:

- Damage to blood vessels.
- Problems with metabolism.
- "Thick blood" (an abnormal tendency to form clots).
- Interaction of free radicals with neurotransmitters.
- Inadequate irrigation of tissues that had been deprived of oxygen.
- Impaired functioning of arteries.
- Clogging of arteries.

Moreover, unlike ginkgo, some of the drugs in this study have the potential to cause toxicity.

A LOOK AT WHAT'S TO COME

In the chapters that follow, I will tell the amazing story of *Ginkgo biloba* and explore its benefits to your daily health routine, including:

- What ginkgo is.
- Where and how it is cultivated.
- Its history.
- The body's responses to it.
- The differences between drugs and herbs.

- How philosophies of medicine differ.
- The science of ginkgo.
- Ginkgo's actions in the body.
- The key active components of ginkgo.
- How ginkgo moves through the body and effects its healing actions.
- How ginkgo combats the effects of stress and aging on memory and improves brain function in the early stages of Alzheimer's disease.
- Ginkgo's ability to improve the circulation of the blood and foster speedy healing by improving the quality and tone of the blood vessels.
- The benefits of ginkgo to sight and hearing, including relief from ringing in the ears.
- The benefits of ginkgo for sexual function and impotence.
- How ginkgo relieves inflammation due to asthma and allergies.
- How ginkgo relieves leg pains and cramps.
- A survey of ginkgo products and preparations.
- Recommended dosages.
- How ginkgo extracts are prepared.

Throughout the book, we will investigate the elegant, complex design of the human body and all its parts. The human body is probably the most ingenious, effective design that we know of. Certainly, some of its structures, like the ear and the cell, are as elegant as works of art. I will explain the design and action of the body as we go along—whether it is the circulatory system, the makeup of the brain, or the design of the cell. This is not a medical text, but we will look closely at the design and function of the body not only to explain the action of *Ginkgo biloba*, but to inspire us to take care of ourselves so as to live happier, more productive lives.

Whether you decide to read this book from cover to cover, or just to dip into the specific parts of it that interest you with the help of the index, the facts about *Ginkgo biloba* should help you understand its value as a natural healer of the body that may help to relieve a group of conditions, ranging from the merely unpleasant to those that seriously compromise the quality of life. I hope that you will acquire new knowledge and understanding in these pages and that you will know how to make that knowledge work for you. The hallmark of the twenty-first century could well be the use of herbs, rather than drugs, to restore and promote health. Drugs can cure disease, but herbs promote health.

CHAPTER 1

The History of Ginkgo

Just as the ginkgo tree has roots that reach deep into the earth, so the history of ginkgo reaches far back into the distant past.

Herbal medicine may have had its origin when primitive peoples sought to improve themselves by taking into their bodies plants that they hoped would endow them with some of the qualities they saw in the plants themselves—the flexibility of the reed, the regenerative power of grasses, the beauty of the flower. If early people sought a plant that would enable them to live long lives, their search must surely have taken them to the ginkgo tree, for it is older than humankind itself.

In this chapter, we will discuss the "folkloric" use of ginkgo, which dates back thousands of years and which, by and large, has proved to be effective, even in the manner in which the ancients used it.

CLASSIFICATION AND CHARACTERISTICS OF THE GINKGO TREE

The name *ginkgo,* sometimes spelled *gingko, gingo,* and *ginko,* is a Japanese word derived from the Chinese word *yinhsing,* meaning "silver apricot." The ginkgo tree is sometimes called the kew or maidenhair tree, because its leaves resemble the fan-like fronds of some maidenhair ferns. In Asia, it may be called the *yin guo*.

This deciduous tree, whose fan-shaped leaves turn yellow in autumn, is planted widely, not only for its beauty but for its hardiness. It thrives in difficult urban conditions—including drought, smog, frost, and low sunlight—and it seems to flourish on pollution! In New York City, it is the most widely planted tree. Its resistance to parasites, insects, and pollution ensure it of an astonishing natural life span of a thousand years, and it can reach a height as great as 120 feet, with a diameter as great as four feet.

In 1712, Englebert Kaempfer, a surgeon and naturalist, became the first European to encounter the tree when he traveled to Japan with the Dutch East India Company. He gave it the name *ginous*, which was his phonetic spelling of the Japanese name for the tree. In his book *History of Japan*, Kaempfer noted simply that "another sort of nuts called ginous [ginkgo] . . . grow very plentifully on a fine tall tree." Carolus Linnaeus, the great Swedish botanist who was the first to classify and systematize the genera and species of plants and animals and give them universal names in the Latin language, called our subject *Ginkgo biloba*, ("ginkgo with two lobes") because its leaves (lobes) were divided in two by a notch.

THE LONG HISTORY OF THE GINKGO TREE

The history of the ginkgo tree reaches back two million years. It has the distinction of being the oldest existing tree species. It is what the great Charles Darwin called "a living fossil," meaning that it is a plant that evolved into its present state early, and persists today much as it was eons ago.

The Ancient History of the Ginkgo Tree

The tree first appeared on this planet during the Triassic period of the Mesozoic era, some 200 million years ago, when reptiles dominated the land, the air, and the sea, eventually evolving into dinosaurs. Ginkgo is the only living representative of the order *Ginkgoales.* The earliest ginkgo species we know of is *Ginkgo yimaensis.* In the late 1980s, a fossil of this species was found in China. It is similar to *Ginkgo biloba,* except that it has smaller fruits and a leaf with more indentations.

Before the Ice Age, the ginkgo tree was found all over the globe. During the Miocene epoch (7 to 23 million years ago), when the climate in Washington state was subtropical, the ginkgo flourished there, as we can see from their petrified stumps, the main attraction at the Ginkgo Forest State Park. Legend has it that the species survived only because Buddhist monks in China and Japan revered ginkgo as a sacred tree and cultivated it in sacred temple areas. However, ginkgo has recently been found growing in the wild in western China, suggesting that the climate of parts of Asia promoted its survival as much as its revered status did.

The History of Ginkgo in China

Reasoning in a way that must have seemed logical to them, ancient peoples believed that the appearance of a plant illustrated its curative benefits. The earliest records of primitive Chinese people show that when they came to see that thorns could pierce and drain an abscess, they decided that a drink made from boiled thorns could cure swelling. They believed that yellow bark cured jaundice, which is a yellowing of the skin and eyeballs. Knowing that bats could fly in the dark, they believed that the feces of the flying rodents could cure blindness. If a person had a rash or sore, it made sense to them to rub snake skins on the affected area because they presumed that the snakes must have shed their skin because it had been bothering them in the way that a rash or sore would.

The Chinese were not alone in these primitive medical practices. Egyptian and Babylonian herbalists used the lungs of the fox to treat chronic breathing problems, and they used cashew nuts, which are kidney-shaped, to treat diseases of the kidneys. Early Native Americans used the tendrils of squash plants to eliminate intestinal worms because the tendrils resembled the worms.

Over the centuries, through trial and error, the Chinese abandoned treatments that proved not to be effective, despite the appearance of the remedies, and developed valuable and demonstrable knowledge of remedies that actually did seem to be effective. This knowledge was passed down from generation to generation. Usually it was passed on orally, but archaeologists have uncovered some evidence of early prescriptions that were scratched on turtle shells and ox bones. Later records were carved on stone. Human knowledge of herbal treatments increased when writing developed in various civilizations. Then prescriptions and remedies were recorded, and, over successive generations, herbalists could see which practices and medications worked, and which did not.

We know that two and three centuries before the birth of Christ, trade brought the Chinese into contact with Indian, or Ayurvedic, medicine, because plants native to India are mentioned in written herbals of the period. However, we do not know what the results of this contact were and we cannot be sure what its significance to the historical development of herbal medicine in China might have been.

The early Chinese sought out edible plants and observed the effects they had on the body. Some helped or cured the body, some caused vomiting, and some caused pain. The genius of these prehistoric herbal scientists was that they did not completely reject the poisonous ones. They reasoned that if a plant had a powerful effect, even if it was a negative one, it had a potency that might have some benefit. In other words, even if its strength caused harm in some instances, there might be other instances in which the power could cure. These early Chinese learned how to use plants like ginkgo to good effect, finding what parts could bring health to the body, in what ways, and in what quantities or doses. In the case of ginkgo, for example, the most obvious thing about the tree is that its fruit causes skin rashes, yet 5,000 years ago, the Chinese learned to avoid this problem while making use of the seeds (also known as ginkgo nuts) and leaves as medications. Through trial and error, ancient Chinese herbalists learned how to use the leaves and seeds of the ginkgo tree as a medication and even as a gourmet treat.

The Chinese hold that three emperors gave herbal knowledge to the people. The first was Fu Si (2953–2838 B.C.E.), a legendary figure who is considered to be the inventor of the eight trigrams (linear signs) that form the basis for the *I Ching, The Book of Changes*. This book introduced the concept of yin and yang. This is the idea that everything—the universe, the earth, and humanity itself—is in the grip of balancing forces, such as light and darkness, damp and dryness, masculine and feminine. These opposite forces grow out of each other. To be healthy, successful and creative, we must seek to keep these opposite forces in balance.

Prizing life and revering the elderly, the Chinese found the most important use of ginkgo very early. The Emperor Shen Nung (2838–2698 B.C.E.), another legendary figure in Chinese history, is considered the first Chinese herbalist and the author of an ancient Chinese medical text, *Pen T'sao Ching*. In it, ginkgo leaves were said to reverse memory loss in the aged in addition to easing breathing problems.

In the case of medical folklore, in keeping with the yin-yang philosophy, a "dry" agent like ginkgo seeds should be taken to counteract "wet" diseases like asthma or chronic diarrhea. Chinese herbalists also used ginkgo to combat chilblains, a swelling of the hands and feet due to exposure to damp cold.

The roasted seeds were considered an aid to digestion and a means of preventing drunkenness. Even today, in Japan and China, the seeds are dyed red and served as a traditional dish at weddings and celebrations where people are inclined to overindulge in food and drink.

Ripe ginkgo fruit, which was prized as a delicacy even long ago in China and Southeast Asia, was soaked in vegetable oil for 100 days and then used as a treatment for pulmonary tuberculosis. The fruit was also believed to aid digestion and expel intestinal worms.

Use of the woody golden ginkgo nut is first recorded in a work entitled the *Household Materia Medica of China* in A.D. 1350. Then, as now, herbalists boiled the nuts to neutralize their toxic elements. However, modern Westerners avoid the nut and use only the leaf to evade the toxicity problem altogether.

The leaves, termed *bai-guo-ye*, are first mentioned in Lan Mao's *Pharmaceutical Natural History of Southern Yunnan*, published in 1436. He recommended their use as external medicine in treating sores and freckles. In 1505, Lui Wen-Tai published *Essentials of the Pharmacopoeia Ranked According to Nature and Efficacy*. In it, he advised that ginkgo leaves should be taken internally to treat diarrhea. Traditional Chinese medicine by this time used *Ginkgo biloba* leaves to benefit the brain and lungs, relieve asthma and cough, and treat diseases brought on by parasites. An infusion of boiled leaves was used to treat frostbite.

Western medicine, introduced to China by missionaries beginning in the sixteenth century, gradually gained favor with the more prosperous people of China. However, the ancient herbal practices survived in rural and poor areas, where these inexpensive remedies continued to be trusted.

With the rise of Mao Tse Tung and the establishment of the People's Republic of China in 1949, the new government sought a means to improve the health of the Chinese masses. Ginkgo's role in herbal medicine was again acknowledged. The modern Western and ancient Chinese traditions worked together. After the Chinese began to admit Western visitors in the 1970s, an American herbal pharmacology delegation studied the 796 prescriptions for folk remedies that were then in use in China and determined that 44.7 percent of them had a basis in fact that justified their use. The group estimated that this was a higher percentage than would result from a similar study of medications used in the United States.

The History of Ginkgo in Korea and Japan

Like Chinese philosophy and written characters, the ginkgo

tree was transplanted to Korea and Japan, where legends about the tree were passed from generation to generation.

The Chinese called the rootlike growths that hang from the ginkgo's branches *zhong ru,* or stalactites. In contrast, the Japanese called them *chichi,* or breasts. In Sendai, Japan, one aged ginkgo that is heavy with these aerial "roots" is said to have been planted over the grave of an Emperor's wet nurse, who promised Buddha that she would give milk to all women who needed it for their infants. The legend is that any woman who prays to the tree will be able to nurse her offspring.

These pendulous structures proved to be the answer to the prayers of gardeners who practiced the Japanese art of bonsai. When the *chichi* were cut from the tree, turned upside down, and planted in pots, they sprouted. Bonsai artists loved to train this unusual plant into unusual shapes, as they still do today.

Ancient Japanese scholars protected their valuable papers from insects by placing ginkgo leaves between the pages of books or near scrolls, and coatings of the seeds have long been used as an insecticide.

Perhaps the most powerful, and touching, demonstration of what the ginkgo tree means in Asia is that ginkgo seeds in the ground survived the blast of the atomic bomb dropped on Hiroshima, Japan, and later sprouted. The living ginkgo survives and triumphs in the park that marks the site of the explosion.

The Discovery of Ginkgo by the West

Europeans first encountered the ginkgo tree in the early eighteenth century (see page 16). They quickly took it home to Europe, where it became popular as an ornamental tree, particularly in urban areas. In 1784, soon after the American Revolution, the tree was brought to North America and planted in the garden of William Hamilton at what is now Woodlawn Cemetery in West Philadelphia.

The ginkgo tree became common enough to find its way

into popular Western literature. Sir Arthur Conan Doyle, creator of Sherlock Holmes, wrote descriptively of it in his science fantasy *The Lost World*. He said, "One huge ginkgo tree, topping all the others, shot its great limbs and maidenhair foliage over the fort, which we had constructed." Oliver Wendell Holmes mentioned it in *The Autocrat of the Breakfast Table* in 1891. In setting a scene, he said, "One of the long granite blocks used as seats was hard by, the one you may still see close by the Ginkgo-tree."

The Modern History of Ginkgo

Ginkgo's recent history is noteworthy. A new chapter in medicine is opening, as traditional, proven "folk" treatments, like *Ginkgo biloba,* become acknowledged as a valid branch of medical treatment. Europeans in the twentieth century have accepted ginkgo's medicinal value in improving mental performance and blood circulation. It has become one of the most popular and respected medications in Europe, where *Ginkgo biloba* extract (GBE) is the herbal medicine most frequently prescribed by physicians and pharmacists. Although Europeans have accepted *Ginkgo biloba* for years, only now is the American medical establishment taking a look at this herb. Certainly herbs have found favor with consumers.

Modern science has proved that the ginkgo tree contains a disinfectant called 2-hexenal, which kills microbes. In addition, the acid on the leaves is poisonous to insects, to the *Escherichia coli (E. coli)* bacteria that have sickened and even killed people in the United States in recent years, and to the mosaic viruses that affect tobacco and beans. So the Japanese ancients knew what they were doing when they planted ginkgo around bean plants to protect them.

"Quaint" practices involving the use of ginkgo for health now make sense. Today in New York City in the autumn, Chinese women go to Central Park to gather ginkgo seeds that have fallen from the trees. They wear gloves to protect themselves from the foul-smelling, skin-irritating fruit as they gather the plum-shaped yellowish seeds, which they add to their soups. They crack and boil the raw seeds until they are soft enough to be eaten on their own. Ginkgo seeds are also sold in cans in Chinese grocery stores under the name of "white nut." Because ginkgo seed has an astringent property, Chinese-Americans consume ginkgo to combat such conditions as asthma and chronic diarrhea.

Today in the United States, herbal products are classified as foods, not as medicines. However, alternative medical procedures are beginning to win respect from established institutions. As I noted in the Introduction, the Office of Alternative Medicine of the National Institutes of Health was established to "more adequately explore unconventional medical practices." The leading health-care network in the United States has announced that patients who seek alternative medical treatments will be reimbursed as readily as those who go to conventional physicians.

Modern pharmacologists are making the most of the benefits of *Ginkgo biloba.* They have taken the study of the benefits of plants beyond guesswork. Scientists employing the scientific method of observation and experimentation are now seeking to find out what works and how.

THE PHYSIOLOGY OF THE GINKGO TREE

The ginkgo tree is a marvel of nature. Its unique properties have enabled it to survived for millions of years, since the heyday of the dinosaurs. The same properties that enabled the ginkgo tree to survive over all that time provide medicine to humankind.

Dr. Del Tredici, a botanist who specializes in the study of the ginkgo tree, believes that the tree's remarkable longevity is due to the fact that it is constantly sprouting. Dr. Tredici observed these sprouts at the bases of the trunks of ginkgo trees in the Tian My Shan Nature Reserve in China. The trees had grown in the wild for at least several hundred years. Dr. Tredici believes that the sprouting behavior is responsible for the fact that many ginkgos have more than one tree trunk. When a ginkgo tree falls, tiny sprouts continue to grow up from the root system. "It has this ability to persist indefinitely by sending up shoots even after the original trunk falls down," Dr. Tredici observes. "When some disaster comes along and wipes out the forest, lo and behold, the ginkgo is ready to spring back up."

Dr. Tredici has also noted that the tree seems to thrive on abuse. At a thousand-acre ginkgo plantation in Sumter, South Carolina, where 10 million trees are grown for their extract, the leaves are stripped at the end of June and a chipper is run over the branches so that the trees will remain short enough to be easily stripped the next year. "They sprout the following spring, which shows how hard we can push these plants without killing them," he says.

The tree has the natural constitution to give itself a long life, and it yields compounds that can improve human health. These compounds, known as flavonoids and terpenes, keep human blood vessels—the veins, arteries and capillaries—sound and also scavenge free radicals, as we shall see. Their wide-ranging activity gives *Ginkgo biloba* its beneficial effects.

SUMMARY

People have been using *Ginkgo biloba* for thousands of years, drawing on its history of demonstrated effectiveness in preventing *and* relieving medical ills. It is probable that in years to come, more Americans will learn to respect herbal treatments and extracts, including *Ginkgo biloba,* which has already won widespread acceptance in other parts of the world. We turn to that story next.

CHAPTER 2

Ginkgo in the Medical Culture Wars

In this chapter, we will discuss the growing importance of alternative medicine, particularly herbal treatments. The use of herbs is the oldest and most common aspect of medicine. Every culture has a history of using herbs for health. Cave dwellers used herbs, and, as we shall see, in the world's most prestigious laboratories modern scientists are studying the active ingredients of ancient herbs and finding benefits that are as encouraging as they are startling. We will look at two cultures—the Chinese and the Indian—in which herbal medicines were employed to maintain a balance of bodily forces considered essential to health. We will also consider the degree to which Europeans have accepted *Ginkgo biloba* and consider why the American medical establishment scoffed at herbal treatments until recently.

CHINESE MEDICINE

The Chinese medical tradition is based on certain philosophical principles. The hallmark of traditional Oriental medicine is that it seeks to balance *chi*, or vital energy. In this tradition, balancing *chi* means keeping energy flowing through the body and preventing the blockages of energy that are believed to lead to illness. Diagnosis involves observing the patient, listening to the patient's complaints, and examining the patient by touch, including feeling the pulses. The Chinese believe that each person has fourteen pulses, seven on each wrist. While the only pulse Western doctors use is the one related to the heartbeat, the Chinese pulses are found by pressing the radial artery by the wrist bone. Each one indicates the state of a specific organ, including the gallbladder, the small intestine, and the sex organs. Supposedly, these pulses are energy paths, lines of energy that regulate the functions of the body. Acupuncturists call them meridians. There are no correspondences for these meridians in Western medicine.

Some 5,000 years ago, the Chinese recognized psychosomatic illnesses, those brought on by mental attitude. One commentator wrote: "If psychic disturbances, caused by emotional shocks, persist, no treatment can be given, because the mind must be at peace in order for the defensive energy to function properly." He also recognized that a patient's economic status could be important in predicting the nature of a patient's illness: "Before examining a patient, we must first learn if he is rich or poor. If he is rich, most of his illnesses will be organic; if he is poor, he is probably undernourished and his defensive energy may be weakened."

The Chinese hold that each individual is a microcosm, or miniature version, of the entire universe because all are subject to the same laws. Each person, and the world, is a blend of *yang,* which means light, and *yin,* which means darkness or shadow. Each exists only in relation and comparison to the other. Yang and yin can also be thought of as heat and cold, wet and dry, life and death, masculine and feminine. All living things embody both principles, and one leads to the other in an endless ebb and flow. Because of these ideas, even ancient Chinese believed that each person combined traits that were considered feminine and masculine. The concept of yin and yang was introduced in the *I Ching, The Book of Changes,* which is traditionally considered to be based on the work of the emperor Fu Si (see page 19).

In the traditional Chinese system, still in use today, each person's physical condition is seen as a manifestation of one of five elements: wood, fire, earth, metal and water. Since the element of fire governs the heart, which controls circulation, ginkgo would probably belong to the category of herbal remedies that promote the proper influence of this element.

The ancient Chinese medical text *Pen T'sao Ching* lists 365 drugs; of them, 252 are of plant origin, 67 derived from animals, and 46 from minerals. Tradition holds that the emperor Shen Nung wrote this text after having sampled all the herbs, including the poisonous ones. However, it is more likely that medical scholars produced this and the other medical works attributed to the emperors.

Huang Di (2697–2595 B.C.E.), known as the "Yellow Emperor," is said to have written *The Yellow Emperor's Inner Classic,* a summary of Chinese medicine. Some 700 years after the birth of Christ, *Tang Materia Medica* listed 844 drugs in use in Chinese medicine at that time. This work comprised fifty-four volumes that described plants, their flavors, and their effects on the body. In 1590, this work was replaced by *Compendium of Materia Medica,* which listed 1,892 drugs, mainly of plant origin.

As described in Chapter 1, Chinese herbal medicine began to decline in prestige after the introduction of Western medical thought, starting in around 1644. However, after the establishment of the People's Republic of China in 1949, Chinese medicine was again taken seriously, and the government encouraged practitioners of modern and ancient traditions to work together.

INDIAN MEDICINE

Chinese medicine is the oldest surviving medical tradition. It is rivaled only by India's heritage of Ayurveda. Those who want to understand alternative medicines should have a basic grasp of Ayurveda, India's traditional medical system. The word *Ayurveda,* meaning "knowledge of life," is derived from the Sanskrit words *ayus,* which means "life," and *veda,* which means "knowledge." The notion of balance and imbalance in health and illness is an ancient and pervasive one that we find not only in the Chinese tradition but in the Indian tradition as well.

Whereas Chinese medicine says that the basis of disease is an imbalance of energy, ancient Indian medical wisdom says that the cause of disease is an imbalance (or stress) in the patient's consciousness. Ayurveda has been practiced for 5,000 years. As China has given the West acupuncture, India has given us meditation, yoga, and biofeedback, which rely on the power of the mind.

To be healthy, according to the precepts of the Ayurveda, each individual must be in harmony with himself and with the society of which he is a part. Mental stress leads a person to act in an unhealthy way, which leads to illness. Not surprisingly, Indians developed yoga and meditative practices to help them restore balance and health to their minds and bodies.

Ayurveda holds that there are three physiological types—vata, pitta, and kapha. Each type has physical characteristics, but these guidelines are less reliable than the psychological traits of each type. Vata persons are quick, erratic, adaptable, nervous, and sensitive. They sleep lightly and do not perspire. They correspond to the element of air, and have "nervous" complaints, stomach trouble, arthritis, and many kinds of vague pain.

The pitta type has a strong appetite and loose, gangly motions. People of this type sleep soundly, sweat profusely, and are easily angered. "Fiery" in disposition, they are prone to inflammations, infections, and liver problems. The soothing qualities of ginkgo are especially good for them.

The kapha type is calm and sentimental, with steady habits, dependable digestion, and a tendency to have mucus. These people are governed by the element of earth and are prone to bronchial ailments, swollen glands, and stomach problems.

While a person of any physiological type may suffer from any disease, the person's type will determine the quality and seriousness of the ailment. Diarrhea may be the most obvious example. The vata type will have gas, pain, bloating, and limited urination. The pitta will have a violent, fevered diarrhea. The kapha will feel heavy and weakened.

In Ayurvedic practice, each type should be brought into balance through herbs and lifestyle changes. Followers of Ayurveda assert that good health rests on three pillars:

- A good mental outlook.
- A healthful lifestyle.
- An understanding of one's physical type.

In addition, to be healthy, one must cleanse the body of toxins and impurities that build up.

Like the Chinese, the Indians have a tradition of using herbs. Like Chinese herbs, these plants are winning the respect of the West.

EUROPEAN MEDICINE

When it comes to the conventional medical establishment, European physicians have always been more open to herbs than their American counterparts. In the United States today, federal law prohibits purveyors of herbal products from marketing them as medicines. Instead, herbal products are sold as food products. No specific disease-related claims can be made on the packaging of herbs or vitamins, unless they have received drug approval from the Food and Drug Administration.

Manufacturers of herbal products have an easier time getting herbs certified for medical use in Europe. The process of getting a natural remedy approved is quicker and cheaper there, as long as the remedy has a track record of safety, as *Ginkgo biloba* does. European regulators rely on what they call the doctrine of reasonable certainty. The European Economic Community (EEC) saw a need to establish overall standards for herbal medicines and created a series of guidelines called The Quality of Herbal Remedies, which is based on Guidelines for the Assessment of Herbal Medicines, developed by the World Health Organization (WHO). Basically, these guidelines say that if a product has a history of safe, effective use, it can be regarded as a safe and valid as a treatment, unless there is evidence to the contrary.

By no means do Europeans rely on the history of an herb alone. In Europe, herbal remedies are classified in three categories:

- Prescribed herbs.
- Over-the-counter herbs.
- Herbal "folk" remedies.

Herbs given by prescription are controlled and documented with great vigilance. Over-the-counter preparations based on herbs are as carefully monitored as over-the-counter drugs in the United States. Herbal "folk" remedies that can be found in health stores and ethnic groceries are judged safe on the basis of their records, but have had less clinical testing.

In Europe, *Ginkgo biloba* has met with approval and has found its way into retail stores selling health products and medicines. Ginkgo has won acceptance because of its demonstrated efficacy in improving circulation, among other benefits, but the extract was also subjected to rigorous testing in laboratories to meet the standards of "phytomedicines" (plant-based medicines) that are given to treat serious conditions.

In Europe, phytomedicines are carefully studied in university and hospital laboratories. *Alternative Medicine: Expanding Medical Horizons,* the report on alternative medicine that was presented to the National Institutes of Health, noted that "in Europe there have been credible research studies reporting positive effects on a variety of chronic illnesses for herbs such as *Silybum marianum* (milk thistle), *Ginkgo biloba* (ginkgo), *Vaccinium myrtillus* (bilberry extract), and *Ilex guayusa* (guayusa)."

In accordance with European guidelines, *Ginkgo biloba* is regarded as a safe, effective treatment. In Europe, doctors prescribe *Ginkgo biloba* extract for problems of "cerebral insufficiency." As an herbal, alternative medicine, *Ginkgo biloba* has given relief to people suffering from a number of conditions associated with age-related cerebral insufficiency, including ringing in the ears, the earliest symptoms of Alzheimer's disease, asthma, allergies, memory loss, the effects of stroke, and leg pains and cramps. *Ginkgo biloba* extracts are among the most commonly prescribed drugs in France and Germany, where the extract is taken orally and sometimes by injection. *Ginkgo biloba* is registered as a drug in both countries for the treatment of organic brain disorders, and the extract is even more widely available over the counter, at concentrations of 40 milligrams per tablet.

Ginkgo in France

In France, herbal products like ginkgo extracts are licensed by the French Licensing Committee and approved by the French Pharmacopoeia Committee. They are labeled "traditionally used for . . ." (the specific benefit), so that purchasers know that the product has not necessarily been tested scientifically but has a history of use. The French recognize the traditional use of *Ginkgo biloba* to stimulate memory, but they have done scientific research on the product as well. It accounts for 4 percent of all prescription medicines sold in France.

Ginkgo in Germany

In Germany, ginkgo is licensed for the treatment of "cerebral insufficiency." This term refers to a host of problems, including impaired memory, dizziness, ringing in the ears, headaches, and nervousness and anxiety. The extract has also been approved as a supplemental treatment for certain kinds of hearing loss and for leg cramps and numbness due to poor circulation.

German regulators consider an herbal product in total, as one active ingredient. This makes their approval process quicker and easier than American drug regulation, which requires that a drug be broken down into its chemical properties and its active ingredient isolated. In the case of extracts like *Ginkgo biloba*, the German Federal Health Office regulates the extract so that its strength and manufacturing process is standardized. They use a "monograph system," which is a review of scientific studies done on a product. The Ministry of Health Committee for Herbal Remedies gathers all the available studies performed by scientists in universities and medical centers. It then reviews them, and once the committee is satisfied that the herbal extract has been proven beneficial and free of harmful reactions, it approves the extract. *Ginkgo biloba* accounts for 1 percent of all prescription medicines sold in Germany, with total sales of $280 million annually.

Ginkgo in England

England relies on the rule of prior use when certifying a product as safe for consumption. If a product has been used for hundreds of years and seems to have positive effects and no detrimental ones, it can be sold as a safe product. To safeguard the validity of this process, the Ministry of Agriculture, Fisheries, and Food maintains with the Department of Health a database of harmful effects reportedly caused by the use of "non-conventional" or alternative medicines.

AMERICAN MEDICINE

The introduction of European medicine to the Americas began when Europeans themselves arrived on this continent. When the Pilgrims set foot on Plymouth Rock, they brought knowledge of English herbs with them from the old country, and they learned as much as they could from the ways that Native Americans used the plants that were native to these shores.

On the Pacific coast, Spanish explorers and missionaries added their medical knowledge to what they learned from the peoples of Mexico and the rest of Central and South

America. When the Chinese began to arrive in California and other points on the West Coast, they contributed their age-old practices to the various strains of what was developing into American medical knowledge. Surely the Chinese who reached America were glad to find ginkgo trees, since they knew that the seeds and leaves of this ancient tree were one of their most trusted herbal remedies.

The Growth of Biomedicine

In the middle of the nineteenth century, Western medicine moved from the herb garden to the laboratory. Medicine became a science, systematic in its approach. Louis Pasteur demonstrated that germs cause disease. Researchers learned that antitoxins and vaccines could prevent horrible diseases, such as tetanus, cholera, and diphtheria. They discovered that proper sterilization could kill bacteria and prevent the ghastly infections that often followed even successful operations. They found that anesthesia could end the excruciating pain of surgery. Armed with scientific knowledge, they developed drugs, including penicillin and the tetracyclines, that saved millions of lives. A drug, according to the definition in the American Medical Association's *Encyclopedia of Medicine*, is "a chemical substance that alters the function of one or more body organs or changes the process of a disease."

"Biomedicine" altered forever the way medicine and health and disease are understood. It continues to advance. Sometimes it results in developments, such as test-tube conception and life-support systems, that have serious moral consequences and leave us worrying over ethical questions. Sometimes biomedicine makes a quantum advance—such as the development of the polio vaccine—that reduces the physical and emotional suffering of humankind. As it advances, biomedicine becomes increasingly expensive, which means that many who need it most—the old, the young, the poor, and those with chronic health problems—cannot make use of it.

Even in a book promoting herbal medicine, I need to acknowledge that modern medicine has given the world incomparable blessings. The public health of entire societies and the private, personal well-being of individuals and those who love them have benefitted from modern medical developments. However, modern medicine's ascendancy led to an environment in which much traditional knowledge has been ignored. Although the traditional use of ginkgo to restore and promote health is centuries old, until recently most American doctors regarded ginkgo as a folk remedy that did not deserve serious medical consideration. America has been anything but a medical melting pot. Scientists looked down on herbalists.

The Decline of Herbal Medicine

Because of its great, nearly miraculous advances, Western scientific biomedicine became the standard of health care. Until the middle of the twentieth century, however, many average Americans continued to rely on traditional home remedies—such as chamomile tea for nerves and mint tea for digestion—to take care of minor health problems. The family doctor was a partner in health, not the sole authority. Physicians themselves often recommended herbal preparations. Until the 1940s, their medical-school textbooks described plants in terms of their medicinal properties. Books of pharmacognosy, or "drug knowledge," contained information on useful tree barks and twigs, plant roots, flowers, berries, and leaves.

After World War II, as medical technology advanced beyond the understanding of most people, Americans lost touch with home remedies, including herbs. Each ethnic group looked with some suspicion and scorn on the medicine of the others—unless they were forced by an emergency to try unfamiliar remedies, much as James Reston was on the occasion described in the Introduction. Often the doubt and wariness were generational as much as cultural. American-born generations of immigrant families disdained the medical wisdom of "the old country" of their forebears. For example, many younger or American-born Chinese think that Chinese medicine is primitive, superstitious, and outdated. In fact, all systems have at least a little wisdom or insight to impart. Selecting from them is a matter of determining which claims for the medicinal properties of an herb or other traditional medicine have a basis in fact, as the American herbal pharmacology delegation did when studying herbal medications in China (see page 20).

Many people do not know that common medicines have a basis in herbs. For example:

- Aspirin is derived from the bark of willow trees.
- Some cough drops are based on horehound and other members of the mint family.
- Lemon is found in many over-the-counter remedies because its pulp lowers blood cholesterol, not just because it tastes good.
- Many ache-easing products are based on camphor—an evergreen tree in the laurel family—and on eucalyptus—an evergreen tree in the myrtle family.

Current Attitudes Toward Non-Conventional Approaches

The American Medical Association's *Encyclopedia of Medicine* defines medicine as "the study of human diseases, including their causes, frequency, treatment, and prevention." It continues, "The term is also applied to any substance prescribed to treat illness."

In the United States, we call non-conventional approaches to medicine—such as acupuncture, chiropractic, and herbs—alternative medicine. In Europe, these are called complementary medicine.

Biomedicine is the primary health-care method for only 10 to 30 percent of the world's people, even today. The rest rely on herbal medicine or some kind of folk medicine that they can afford and that they consider effective for them. The

World Health Organization (WHO) estimates that 80 percent of the world's population, or 4 billion people, currently use herbal medicine as their primary method of health care. Most herbal medicine practices give equal emphasis to the prevention of physical ailments and the cure of existing ailments. Herbal products are relatively inexpensive, and can safely be used for long periods of time.

Meanwhile, in developed Western countries, more and more people are discovering the unpleasant side effects of pharmaceutical drugs. The wholesale use of antibiotics has led to yeast infections by destroying the balance between "friendly" bacteria killed by antibiotics and yeasts and fungi resistant to antibiotics. As a consequence, yeasts can proliferate when their natural enemies—the bacteria—have been killed by antibiotics. Overuse of antibiotics can also lead to darkened teeth as a result of vitamin B deficiency, which can occur if useful intestinal flora are destroyed and yeasts and fungi proliferate. Many people have become wary of using such strong drugs for common ailments and fear that they will become overmedicated. They are also reluctant to pay the great financial price of most modern drugs. They see herbs as cheaper, less toxic alternatives to drugs. Many people are also choosing to avoid so-called "heroic measures," such as chemotherapy, because of their fear of side effects.

Although herbs are clearly the basis for many medications, in the United States today, herbal products and vitamins are marketed not as medicines, but as food products. Unless an herb has received approval from the Food and Drug Administration, no health claim can be made for it. We would probably be better off if herbs were recognized and regulated, but the FDA approval process is a long one, requiring an involved series of laboratory tests and clinical trials on animal and human subjects. The cost of this large-scale testing is such that it requires corporate sponsorship, usually from a drug company seeking to bring a product to market, and since herbs are natural substances that cannot be patented, there is little incentive for drug companies to undertake this.

However, so many people have found relief for their physical ailments at local stores where herbal products are sold that there is a new and growing determination to rediscover and promote the benefits of herbs. In large American cities, stores selling Chinese, Indian, and Latin American remedies can be found in many ethnic neighborhoods.

As more exotic foreign foods—from curries to dim sum—become familiar, Americans are learning more about exotic herbs and spices, which opens their minds to the idea of using medicinal plants. Many of us feel increasingly willing and competent to explore the benefits of herbal treatments—particularly after being disappointed by prescription drugs for such ailments as arthritis, AIDS, and asthma.

It is probable that in years to come, Americans will learn to have more respect for herbal treatments and extracts. *Ginkgo biloba* will probably be among the most popular of these. Its value and efficacy have already been established and acknowledged in Europe. In the last fifteen years, interest in the medical value of ginkgo has been growing dramatically. Word has spread of the herb's astonishing ability to improve circulation, particularly in the central nervous system, which includes the brain and spinal cord. *Ginkgo biloba* is especially helpful in improving the functioning of the brain and the spinal cord after injury.

SUMMARY

We have discussed the fact that although the Chinese have counted on the healing properties of *Ginkgo biloba* for centuries, the herbal product has had a difficult time breaking into the American medical scene. Modern Western medicine brings great benefits to humankind, particularly in cases of traumatic injury and infectious diseases. The danger is that it can be captivated by technological developments and caught up in impersonal procedures. Finally, however, Western medicine has turned its research tools on herbs like *Ginkgo biloba.* Europeans, particularly the French and Germans, acknowledge its proven history as a boon to humankind and have welcomed it as an over-the-counter product. Scientists in Europe have studied it in laboratories, so that it is also marketed as a prescription drug. So we see that the test tube is validating the herbal shelf. Much of the herbal knowledge of the Chinese medical system and the Indian system of Ayurveda have been confirmed through practice and examination in the laboratory. Knowledge of the medical benefits of herbs would have been lost if herbs had not been valued and their medical benefits recorded throughout history. Scientific knowledge has confirmed the wisdom of ancient peoples and their herbal, alternative medical traditions, as we are about to see in the next chapter, on the science of ginkgo.

CHAPTER 3

The Science of Ginkgo

Thousands of years ago, Asians demonstrated that *Ginkgo biloba* helped humankind by improving circulation and memory and relieving respiratory problems. Modern scientists accepted the fact that ginkgo had been used for thousands of years, but wanted to know why and how it worked. The scientific method developed in seventeenth-century Europe consists of, first, stating a question or problem to be resolved, and then suggesting a possible solution. Through experiments one learns if the proposed suggestion works. Then the results of those trials are interpreted and sometimes a new theory is devised.

There are two questions scientists have asked about *Ginkgo biloba*:

1. Does ginkgo work to relieve specific problems?
2. Is ginkgo better at relieving these problems than other substances available on the market?

In this chapter we are going to learn to understand the science of *Ginkgo biloba* better. We will consider scientific research that points toward the reasons for *Ginkgo biloba's* benefits to so many areas of the body, from the eyes to the lower limbs, from the brain to the most microscopic capillary. Scientists asked if *Ginkgo biloba* would work to improve brain function, leg cramps, and the health of the retina of the eye. They tested it in the lab and on living patients, and found that in most cases *Ginkgo biloba* helped these and other areas of the body.

Although the working of ginkgo is still not totally understood, its many healing properties are attributed to the complex structure of its molecules and particularly to certain types of compounds, known as flavone glycosides, or flavonoids, and terpene lactones, that it contains. The West began to standardize the plant's properties in the late 1950s, extracting the key active components, whose combined actions account for ginkgo's antioxidant activity. The terpene lactones fall into two categories, designated ginkgolides and bilobalides. These compounds, which improve circulation and stimulate regeneration of nerve cells, are unique to this herb.

THE SCIENTIFIC STUDY OF HERBAL MEDICINES

Scientific knowledge in the ancient Orient was based primarily on observation rather than on systematic experiments. The only important thing to ancient investigators was whether or not an herb reversed an unhealthy condition. Once they knew that it worked, ancient Chinese investigators—and ancient Egyptians and Mesopotamians, for that matter—did not invest more of their time wondering about the reasons why it was so. They were satisfied if their headaches, toothaches, and eyestrain were relieved, and they attributed the results to what we might call the Spirit of the Plant.

However, the Greeks, who were in many ways the source in Western civilization, were more interested in impersonal reason than in the spirit. They looked for underlying causes in the world around us. Reasoning alone did not always lead them to the correct answers, any more than trial and error had led the Chinese herbalists to immediate cures. For example, the Greek philosopher Thales, who was the first to investigate natural phenomena beyond the mythological explanations for them, had many successes. He studied astronomy carefully enough to predict an eclipse of the sun in 585 B.C.E.—but he also believed that the earth was a flat disk floating on a sea of water.

Modern Western science did not develop until the seventeenth century, when its practitioners began to follow Galileo Galilei's lead in insisting that theories be tested through experimentation, and that explanations of phenomena be verified through research using newly developed instruments, such as the telescope and microscope. Through these methods of experimentation and verification, Europeans learned not only that certain things worked, but often they worked. Their new knowledge of causes and effects often led to other discoveries and applications. For example, Gregor Mendel wanted to understand why living things existed in such variety. Through experimentation, he demonstrated the existence of dominant and recessive genes. His work on inheritance factors did not create much of a stir at the time—in fact, it was virtually ignored for fifty years after its publication in 1866—but it led the way to the modern study of genetics, including our understanding that a predisposition to certain diseases, such as Alzheimer's disease or hardening of the arteries, may run in families.

In the Western world, the first scientist who was drawn to study the basis for the ways plants benefitted humankind was a Swiss named Philippus Aureolus Theophrastus Bombast von Hohenheim (1493–1541). Working under the name of Paracelsus, he opposed the existing theory of the "humors," the belief that diseases were the result of imbalances of bodily fluids. Instead, he theorized that each disease was caused by a specific agent from outside the body, and that therefore each disease could be treated by ingesting another agent, a specific substance that would help the body combat the disease. He was considered so arrogant that the word *bombastic* has been said to come from his name, but his quest in life led to medicine becoming a scientific, rational process, rather than an oral, folkloric tradition based on random trial and error.

Of course, the Chinese had been working on these problems since around 2500 B.C.E., and Paracelsus didn't come along until the fifteenth century, during the Renaissance. Since his time, Western scientists have investigated the chemical and biological activities of fewer than 10 percent of the estimated 250,000 to 500,000 plant species. Fortunately, the science of the *Ginkgo biloba* tree has been studied extensively.

THE STRUCTURE OF THE GINKGOLIDE MOLECULES

Interestingly enough, scientists had a difficult time discovering the chemical makeup of the ancient ginkgo tree, and when they did, they did not at first think that it had much use!

This scientific detective story began in 1932, when Japanese scientists working on ginkgo leaves were able to isolate an interesting bitter extract. Over time, German and Japanese chemists divided this bitter extract into four distinct chemical compounds, designated ginkgolides A, B, C, and M.

In 1963, Koji Nakanishi, a Japanese chemist born in Hong Kong in 1925, arrived at Tohoku University in Sendai, Japan, where he planned to study the ginkgolides and to work on discovering their ultimate structure. At about that time, a typhoon damaged many of the ginkgo trees in the city and provided an unexpected lucky break that aided the work of his team. The bark of the roots of the ginkgo tree is the best source of gingkolides, but ordinarily this can not be harvested without destroying the trees. Since so many ginkgo trees had been uprooted by the typhoon, Nakanishi and his team were presented with a natural "harvest" of ginkgo roots. With the help of several workmen, they were able to obtain the roots they needed for their research. Nakanishi and his team made some advances in understanding the structure of the ginkgolides, and he presented their preliminary findings when he gave a paper in Stockholm in 1966.

Meanwhile, another team of Japanese chemists came up with a similar description of the components of ginkgo. The two experimental teams verified each other's work, and the chemical composition of ginkgolides A, B, C, and M were found. In 1966, Nakanishi discovered the "cage" structure of the ginkgolide molecule. This is a unique configuration that at the time of its discovery had never been seen in nature before.

THE ACTIVE INGREDIENTS IN GINKGO BILOBA

The medical benefits of *Ginkgo biloba* depend on the proper balance of its two active components—ginkgo flavone glycosides and terpene lactones. Although the various elements of *Ginkgo biloba* each play a separate role, experimental studies show that the action of all these elements together makes them more effective than any of them would be alone, much in the way that a talented team of football players is more effective than any one talented player, even a quarterback, could be alone.

The relative amounts of these two components in *Ginkgo biloba* were determined in the course of a study commissioned by the German government prior to its approval of *Ginkgo biloba* for sale over the counter. The German Federal Health Agency set up a commission composed of doctors, druggists, toxicologists, and others knowledgeable in developments in medicinal plants to review herbs and herbal combinations. The commission approved 200 of them for sale over the counter, including, not surprisingly, *Ginkgo biloba* extract. The commission published a detailed monograph listing ginkgo's benefits and composition and the specifications for manufacturing the extract properly under quality-control standards. The monograph described the active components of *Ginkgo biloba* as:

- Flavone glycosides (22 to 27 percent).
- Terpene lactones (5 to 7 percent).

Check for proper levels of both of these compounds when you evaluate the *Ginkgo biloba* extracts on your store's shelves. You might also want to know what these forbidding chemical names mean.

Flavone Glycosides

Flavone glycosides are a type of flavonoid. The flavonoids are a group of substances found in many plants and fruits, particularly citrus fruits. They are antioxidants, which means that they "clean up" or "scavenge" free radicals. Flavonoids also protect the cells against the breakdown of arachidonic acid, an unsaturated fatty acid that keeps cell membranes healthy and permeable. The flavone glycosides in *Ginkgo biloba* include the compounds quercetin, kaempferol, and isorhamnetin.

As a group, the flavonoids are responsible for the antioxidant activity of ginkgo. *Ginkgo biloba* extract, taken for fourteen days, may be more effective than beta-carotene (provitamin A) and vitamin E as an antioxidant. In addition to acting as antioxidants, by reducing the "stickiness" of the platelets in the blood—that is, their tendency to adhere to one another and thereby coagulate the blood—the flavone glycosides improve circulation and prevent the formation of clots in blood vessels. Further, experiments indicate that flavonoids make vitamin C more effective, so that the body requires lesser amounts of it. *Ginkgo biloba* saves and protects vitamin C. You can get the effect of large doses of vitamin C if you take smaller doses with ginkgo.

Bioflavonoids like those found in *Ginkgo biloba* are known to protect blood vessels. Flavonoids increase the strength of capillary walls, reduce inflammation, and prevent blood cells and proteins from seeping into the tissues. If you take certain steroids for asthma or arthritis, you may bruise easily because those potent drugs make the walls of the blood vessels fragile. However, if you take flavonoids to combat these conditions, your blood vessels should remain strong and able to resist trauma, such as that resulting from the sudden pressure of bumping into something.

Terpene Lactones

Ginkgo biloba extract also contains terpene lactones, in the form of bilobalides and ginkgolides A, B, and C, and ginkgolide J, which German organic chemists isolated in 1987. The terpene lactones deliver another type of necessary help to the body by offsetting processes that lead to the formation of unwanted blood clots.

The terpene lactones in *Ginkgo biloba* improve circulation to the brain and other parts of the body, bring oxygen to the tissues, and enhance glucose uptake—the absorption of glucose (blood sugar) by the body's tissues. Glucose is the body's chief source of energy. Most of it is derived from carbohydrates in the diet, although the interactions of fats and proteins in the cells produce a small amount. By helping the cells metabolize glucose, the terpene lactones in *Ginkgo biloba* promote the body's health and energy.

Terpene lactones also protect nerve cells from damage during even brief periods of oxygen deprivation that can lead to stroke. The terpene lactones account for ginkgo's ability to improve memory and mental function, and can promote recovery from stroke.

Bilobalides and ginkgolides are found only in ginkgo. As we have seen, the molecules of the three ginkgolides A, B, and C have a structure unique in the plant kingdom. They look like cages, and for that reason they are called cage molecules. It is impossible for even the best chemist to synthesize them or to copy them in the lab. To obtain the ginkgolides, you need the ginkgo leaves from nature, rather than a synthetic copy.

THE PRIMARY EFFECTS OF GINKGO BILOBA

To be active in the body, a substance has to bond chemically to receptors on the cells of the body. The term *receptor* refers to a specific structure or area associated with a cell that has a chemical and physical structure that allows it to link up, biochemically speaking, with a particular substance or group of substances. For example, the cells of the heart and lungs have receptors that are able to link to epinephrine, a naturally occurring hormone that increases the heartbeat and sends blood to the muscles. Medicines work because they are able to bind with certain receptors in the body, thereby causing (or in some cases, blocking) a specific biochemical or physiological activity. Particularly in the case of drugs, the effect of a substance meeting a receptor is proportional to the amount of the substance in the region of the receptor. If you don't administer enough of the drug, you get no result. If you give too much, you may get side effects.

In 1900, Nobel prize-winning bacteriologist Paul Ehrlich used the example of the key in the lock, or the nut in the bolt, to explain how medicines find their way to the proper cells in the body. A substance like a drug has a certain configuration and is designed to match a receptor that will accept this physical structure, like a key in its lock. Under certain conditions a receptor will not accept a substance; under other conditions a drug can cause the cell to release its components and degenerate.

Ginkgo does its work not only through interaction with receptors, but also by inhibiting enzymes and other mechanisms that break the body down. The beneficial effects of the extract are due to the combination of its various abilities to protect, to cure, and to regulate activities that occur in certain degenerative processes, including aging.

Improving Circulation

The fundamental reason that *Ginkgo biloba* works as medicine is that it makes the circulatory system more efficient by improving the tone and elasticity of the blood vessels. These include the arteries, which pump oxygen-bearing blood through the body; the veins, which carry carbon dioxide and waste products back to the lungs to be re-oxygenated; and the capillaries, fine blood vessels that distribute the blood to the tissues.

Through its beneficial effect on the circulatory system, *Ginkgo biloba* is able to keep the cells of the body whole and healthy and keep the mind clear. By keeping the circulatory system healthy, *Ginkgo biloba* helps to bring healing, cleansing, enriching oxygen to all parts of the body, particularly the brain, the most important organ of the nervous system, which is the "control center" for the human body. Blood and oxygen flow to the brain is improved, and damaged tissues served by fine capillaries, such as the eyes and ears, can recover at least some of their former health after injury. Cited throughout this book you will find studies conducted on human beings in which *Ginkgo biloba* has increased blood flow to the brain and throughout the nervous system, thereby improving mental activity, including short-term memory and quick-wittedness.

Preventing Undesirable Blood Clotting

Ginkgo biloba, specifically the terpene lactones it contains, counteract platelet-activating factor (PAF). This body chemical is a mediator that initiates or accelerates coagulation or clotting of blood, turning it into a jelly-like "sludge" that does not flow easily. Platelet-activating factor has been implicated in:

- Damage to nerves.
- Disorders of the central nervous system.
- Inadequate flow of blood to the brain.
- Bronchial constriction, which decreases the lungs' ability to take in oxygen.
- Allergic reactions and asthma.
- Inflammation.
- Transplant rejection.
- Shock.
- Irregular heartbeat.
- Kidney disease.
- Chronic problems due to head injuries.

The actions of PAF can ultimately damage nerve cells, slow the flow of blood to the brain, and promote bronchial constriction, decreasing the overall intake of oxygen and leading to a host of problems.

Protecting Against Damage to Blood Vessels

Ginkgo biloba is able to help memory, improve general brain function, and counteract problems with circulation because it benefits the blood vessels and therefore benefits the performance of various organs, most notably the brain.

Ginkgo does several things to aid blood vessels. It maintains their strength because it helps deliver glucose to the cells, which ensures that the blood vessels have the energy they require. It also scavenges free radicals, which can damage the cells and tissues; neutralizes dangerous substances in the blood; and works to open constricted blood vessels.

Significantly, while ginkgo is working to open constricted blood vessels in one part of the body; it does not have negative effects on normal blood vessels elsewhere in the body. In other words, it does not open constricted blood vessels at the expense of opening normal blood vessels too much. It repairs what is damaged and leaves what is working properly alone. That is seldom the case with other medications.

Scavenging Free Radicals

A remarkable effect of *Ginkgo biloba* is that it protects cells by scavenging free radicals, which can cause cells to degenerate, without setting off unwanted chemical reactions. A free radical—sometimes called just a radical—is a highly reactive molecule or atom that is "missing" an electron, leaving it in an unbalanced state. As a result, the free radical can take an electron from another atom or molecule. It is as if the free radical has a hook with which to grab on to a new, and possibly not helpful, partner. Some free radicals are the result of natural biochemical processes in the human body; others are caused by external factors, such as exposure to sunlight or x-rays, radiation therapy, nuclear emissions, or pollutants in the air.

In the human body, free radicals have both beneficial and detrimental effects. The white blood cells of the blood use them to destroy invading organisms, such as bacteria, which is good. However, they also can cause the cells of the tissues to break down, leading to inflammation, such as in arthritis; to heart problems; to Alzheimer's disease; and to other problems. Ultimately, cells can die from the effects of free radicals. In addition, the formation of free radicals can generate a chain reaction, in which one free radical takes an electron from another molecule, leaving that molecule as an electron-deficient free radical that can take an electron from yet another molecule, and so on, producing more and more free radicals that promote unnatural and harmful biochemical processes.

Normally, free radicals are kept under control by free-radical scavengers known as antioxidants, such as the enzyme superoxide dismutase. These scavengers donate some of their own electrons to neutralize the electron-deficient free radicals.

The good news is that *Ginkgo biloba* is an antioxidant, a free-radical scavenger. Its chemistry is such that it can donate electrons from its own molecules to neutralize free radicals. A study done at the University of California–Berkeley reported that "in vitro evidence has been reported that EGb 761 [*Ginkgo biloba* extract] scavenges various reactive oxygen species." Berkeley researchers set about to find out why. In various tests, reported in *Biochemical Pharmacology* in May of 1995, they found that *Ginkgo biloba* was effective in scavenging peroxyl radicals, a type of free radical, which bolstered the theory of the antioxidant therapeutic actions of *Ginkgo biloba*. This antioxidant property is believed to be the basis of ginkgo's many beneficial effects on the brain and circulation, and on the eyes as the ears as well.

Free radicals can promote cancer and aging because they can attack lipids (fatty substances) that are vital constituents of cell membranes, proteins, and nucleic acids (RNA and DNA, which store and transmit coded instructions concerning how cells should function in the body). Lipids in general are very attractive targets for free radicals. When free radicals attack these vital lipids, they damage the cells and in some cases even cause them to mutate. The flavonoids in *Ginkgo biloba* guard against this kind of damage.

The brain and central nervous system have high concentrations of lipids in their cells because lipids contain a great deal of energy, and the nervous system needs a great deal of energy. After all, the nervous system processes all the sensations of sight, smell, hearing, taste, and touch. By protecting the lipids in the brain against damage from free radicals, ginkgo protects the functions of the brain, including everything that make us conscious, as well as subconscious processes we do not even think about.

Protecting Cells from Damage

Ginkgo biloba, carried through the bloodstream, helps to regulate the health of the cells, and *Ginkgo biloba's* diverse effects are probably due to the fact that it helps every cell. Different types of cells in the body vary widely. Some nerve cells, for example, are three feet in length, while mature red blood cells are only a microscopic 0.0003 inches wide. Despite this, most cells are basically similar in structure. Each contains fluid material, called cytoplasm, surrounded by a cell membrane. The membrane, which is made up of lipids and proteins, holds the cell together. It allows necessary substances like oxygen and nutrients to enter while filtering out harmful or inappropriate materials, such as hormones that are meant to act on other kinds of cells. Except for red blood cells, each cell contains a nucleus, which contains genetic coding and functions as a control center, regulating the kinds of enzymes and proteins the cell produces.

Arachidonic acid, a fatty acid, is one of the lipids that forms part of the membrane of every cell we have. If this fatty acid is attacked by free radicals, new chemical compounds result, including prostaglandins and thromboxanes, which promote inflammation that can damage healthy cells. Ginkgo stops the cascade of arachidonic acid derivatives without causing harmful side effects. It thus promotes and regulates the good work of arachidonic acid as a structural part of cell membranes. This has been verified and confirmed by studies reported in the French medical journal *La Presse Médicale,* September 25, 1985.

Combatting Effects of Radiation

Modern science has found solutions to most of the public health problems that plagued the world early in this century. Except in the poorest areas of the globe, diphtheria, polio, and yellow fever are no longer serious health threats. There is even hope that AIDS, a relatively new infectious disease, will be brought under control. However, even as science solved those problems, it created others—among them the deadly threats unleashed by the atomic age. Exposure to radiation causes cells to break down. In fact, we now know that radiation from the sun, which has been with us since the dawn of time, also breaks down our cells and leads to the formation of dangerous free radicals. *Ginkgo biloba* has been shown to be the most effective known antidote to the breakdown of cells due to radiation, so much so that it has helped the victims of the Chernobyl nuclear disaster.

In the blood of people exposed to radiation we find substances known as *clastogenic factors*. These factors can break chromosomes and severely damage cells. The greater the dose of radiation received by a person, the higher the number of clastogenic factors in his or her blood. Survivors of the atomic bombing of Hiroshima and Nagasaki have had these factors in their blood for more than thirty years. Clastogenic factors were also found in thirty-three of forty-seven people who worked to clean up and dismantle the Chernobyl nuclear plant in the former Soviet Union after it malfunctioned in 1987.

In a study reported in *Free Radical Biology of Medicine* in June 1995, researchers studied thirty workers who were exposed to nuclear radiation after the Chernobyl accident and found great damage to their chromosomes, which meant they were at high risk for cancer. Injections of known preventive drugs are dangerous if taken over the long term, so the workers were treated with *Ginkgo biloba*, which has never demonstrated serious side effects.

They each received 40 milligrams of *Ginkgo biloba* three times a day for two months. At the end of that time, the clastogenic activity of their blood plasma—the breakdown of their cells—ended on the first day after the end of the treatment. In follow-up studies, this effect was found to persist for up to seven months, even if the subject did not take any further treatment. Ten months after treatment, one-third of the subjects again showed clastogenic factors. So two-thirds of the workers continued to have *no* clastogenic factors.

Almost as important as the indication that *Ginkgo biloba* combatted the effects of radiation was the news that most people did not have to take it on a continuous basis. Most could receive the benefits by taking it for two months and then not taking it for six or eight months. That shows how potent *Ginkgo biloba* is in scavenging free radicals. There were no side effects, as there would have been with chemotherapy, and it saved money because patients did not need to take it on a daily basis.

I would recommend that anyone who has been exposed to radiation inadvertently, or who will be exposed to radiation as part of medical treatment, take *Ginkgo biloba* to protect him- or herself from the bad side effects, without impairing the good effects. As a preventive medicine, *Ginkgo biloba* also works against sun damage. It is important that people who are outdoors a lot, whether for work or recreation, take ginkgo to protect their skin cells. There is experimental evidence that ginkgo shortens the time skin takes to recover from sunburn and that it reduces the pain of the burn. It also helps to avert long-term damage from the sun, not only to the skin but also to the retinas of the eyes, as we shall see in a later chapter.

SUMMARY

In this chapter on the science of *Ginkgo biloba*, we have seen that this herb, particularly in extract form, benefits the human body in a number of ways, not only as a treatment, but also as a preventive medicine. Ginkgo promotes the health of the blood vessels, thereby improving circulation of the blood throughout the body. Most importantly, it brings refreshing oxygen to the cells and tissues of the brain, thus improving the mind and mental performance. In addition, it scavenges, or neutralizes, free radicals—atoms or molecules that are able to attach themselves to other molecules or atoms and trigger their breakdown and deterioration, which can lead to cancer, Alzheimer's disease, and aging, among other things.

At the beginning of this chapter, we said that there are two questions scientists have asked about *Ginkgo biloba*:

1. Does ginkgo work to relieve specific problems?
2. Is ginkgo better at relieving these problems than other substances available on the market?

Does ginkgo work to relieve specific problems? The answer is yes. We may not know everything about the molecular biology of the different components of *Ginkgo biloba* extract, and we may not fully understand the mechanisms by which these components relieve specific problems, but we do know that these extracts have been helpful to people for problems including various diseases and degeneration due to the aging process. It has a beneficial effect on blood vessels large and small, and it keeps the vital membranes of human cells working normally.

Is ginkgo better at relieving these problems than other substances available on the market? Yes. It compares favorably with most drugs available because it has minimal side effects. Its advantage over other substances comes from the fact that it works on the cell membranes at different levels. It prevents degeneration of the cells and inflammation that directly or indirectly release free radicals and toxins.

The combination of these different actions on so many specific organs of the body is an effect unique to *Ginkgo biloba*.

We shall see in later chapters the specific positive effects that ginkgo has on the eyes, brain, ears, and other body parts. I have not found any other substance that has such varied and far-reaching effects. My bias is that if we were one day to isolate one major substance found in ginkgo, or to test one synthetic substance based on it, and try to substitute that for the extract derived from the natural plant, the attempt would be a failure, because the synergistic effect of all of ginkgo's components, the total effect, is the real one.

CHAPTER 4

Ginkgo and the Brain

The human brain is a three-pound mass of pinkish-gray tissue. Its 10 billion nerve cells, or neurons, manage all mental functions and control movement, sleep, hunger, thirst, and all the instincts and emotions from love to hate to sadness.

Nature gave us an incomparably complex gift in giving us the brain. It makes us who we are and enables us to imagine who we might be; it makes each of us the particular person some people love and others dislike. It is the chief reason for our individuality, but it seems we have to prove ourselves worthy of it by training our brain and taking good care of it.

Ginkgo keeps the cells of the body healthy, including those of the brain, and so it keeps the mind clear. This benefit is not of interest only to people who fear the loss of mental acuity from the effects of aging. Even people who are at the beginning of their adult lives, for whom aging is not yet a concern, want to sharpen their minds to be more competitive in today's demanding social and business climate.

In this chapter, we will take up the structure of the brain, the process of storing and retrieving memories, and the ways in which various conditions of the body—including blood pressure—affect the brain and can be improved by *Ginkgo biloba.*

THE STRUCTURE OF THE BRAIN AND THE NERVOUS SYSTEM

The brain is, in effect, the body's computer center. It takes the vast amount of information supplied by the eyes, ears, tongue, and skin, and processes it for use. All those sense organs improve when *Ginkgo biloba* is taken, as we shall see, and so does the central processor of the information they provide—the brain itself. Before going into specifics about the effect of ginkgo on the brain, let us understand the mighty action of nerves in the body.

The nervous system acts as the body's circuit network, carrying data gathered from the world outside the body (such as the image of an oncoming car) as well as from the inner world of the body itself (such as pain). Its signals alert the brain to the fact that a car is approaching, and then carries the signals from the brain that cause the body to move away from the oncoming car and cause more glucose to be sent to the muscles so that they can act quickly to get the body out of the path of danger.

The nervous system is divided into two parts:

1. The central nervous system, consisting of the brain and the spinal column.
2. The peripheral nervous system, which includes the voluntary system that we control by will (moving our arms, for example) and the involuntary nervous system that is unconscious or automatic (increasing the flow of blood to our arms when we lift heavy weights, for example).

The brain is probably the most complex and versatile device on earth. Some time ago, medical researchers thought that more intelligent people had larger brains, but we now know that that is not so. Intelligence is now considered to be related to the number and type of connections between the neurons and the different parts of the brain and to its complex biochemistry. Somewhere between 100,000 and 1 million chemical reactions occur every minute in the normal brain.

The brain consumes 20 percent of all oxygen taken in through the lungs and distributed throughout the body by the blood. If they are deprived of sufficient oxygen, brain cells can be irreparably damaged, because they do not regenerate themselves as other types of cells do. As I shall discuss in greater detail later, the key to brain function is maintenance—keeping the brain operating properly.

Different areas of the brain control different parts and functions of the body. For example, the medulla, at the base of the brain near the spinal cord, controls the survival functions of breathing, blood circulation, and sleep. The cerebellum, just above the medulla, coordinates body movements. Although the brain has different parts, they do not act independently, but function as a complete whole.

The difference between the brain and the mind has inspired a variety of philosophical, religious, and scientific debates over the centuries. Now the consensus seems to be that the mind is best described as "what the brain does." Studies of brain waves seem to indicate that the right side of the brain controls artistic and imaginative thought processes, while the left controls the more rational and concrete ones.

MEMORY

The human brain seems to be better at storing information than retrieving it. Even so, human memory is more advanced than a computer's memory, however unreliable the human memory can sometimes appear. The physical basis of memory is not yet fully understood.

Looked at in a technological way, memory is the mechanism for storing and accessing data in the brain. Its meaning, of course, transcends that mechanistic description. Memory is essential to our understanding of the world around us, to our making our way through life, and even to our being in love, because it is central to learning, to thinking, and to our ability to function in the world. There are four aspects to memory:

1. Recollection of events based on some kind of reminder.
2. Recall of something that is unaided by any cue.
3. Recognition based on similarity of two events.
4. Relearning, or learning on the basis of having known something before.

Researchers have puzzled for years over how memories are stored in the brain. Some think that certain memories are stored at specific sites; others think that memories are distributed in many areas, and that retrieval depends on the brain's bringing the many "pieces" of a memory together. Both theories may be partially right in the case of human beings.

Research has demonstrated that our brains contain more information than we can access. When neuroscientists stimulate a person's brain electrically, he or she can suddenly recall specific events that had seemed to have been forgotten long ago. The subject's recall is perfect in detail. All the information about an event—the people who were there, what they wore, the sounds around them—is suddenly available to the subject and apparently was there all the time.

The clue or reminder that can bring to mind a memory that had seemed lost is enshrined in literature as Marcel Proust's account of the madeleine, a small cake that made the narrator of his great novel *Remembrance of Things Past* recall a portion of his childhood in perfect detail:

> I raised to my lips a spoonful of the tea in which I had soaked a morsel of the cake. No sooner had the warm liquid, and the crumbs with it, touched my palate than a shudder ran through my whole body, and I stopped, intent upon the extraordinary changes that were taking place. . . . I stop my ears and inhibit all attention to the sounds which come from the next room . . . and I feel something start within me, something that has been embedded like an anchor at a great depth; I do not know yet what it is, but I can feel it mounting slowly; I can measure the resistance. I can hear the echo of great spaces traversed.

Who among us does not find that he or she remembers specific times and people when presented with certain tastes and smells—drying autumn leaves, a special cheese, even the steamy, soapy smell of clothes being washed.

THE ROLE OF THE BRAIN IN OVERALL HEALTH

Even orthodox Western medicine accepts the likelihood that there is a relationship, not yet fully defined, between the health of the mind and the health of the body—*mens sana in corpore sano* (a sound mind in a sound body), as the Romans put it—and that the mind and the body affect each other. At its most basic, this mind-body relationship can mean that people who are told they are sick may begin to feel sick; people who are sick but are told that their laboratory test results have improved may begin to feel better. People are not *made* sick by being told that they are sick, any more than they are made well by being told that they are well; however, they may be made to *feel* that they are sick or to *feel* that they are well. In the last few decades, Western researchers have looked at this mind-body connection in studies of physiology, biofeedback, and the placebo effect.

Indian yogis have known for centuries that by controlling their minds they could control some of their bodily processes. The West began to study the whys of this phenomenon three decades ago, when scientists hooked up electrical equipment to the brains of research subjects and studied biofeedback. In 1968, Herbert Benson, a cardiologist at Harvard University, tested the effect of meditation on high blood pressure and found that meditating lowered blood pressure. He later wrote the popular book *The Relaxation Response* (Avon, 1976) about his findings.

Athletes were among the first people to put biofeedback techniques to work in controlling their bodies. On the tennis court, players were seen talking to themselves. Later, they revealed they were improving their tennis games by giving themselves positive instructions on how to win.

It took modern Western medicine some time to see that emotions affect the body. However, experimental scientists have long acknowledged that a subject's psychological state can skew the most rational experiment. One example of the effect of psychological factors on the body, or the mind-body phenomenon, in experimentation is the placebo effect. When researchers study the effect of medication on living patients, they typically divide the population of patients into two groups, an experimental group and a control group. The experimental group is given the medication that the researchers want to study, and the other group of subjects is given a placebo, a pill or dosage that contains no medication. In essence, the placebo is a fake given to equalize the perception of the patients so that their minds cannot affect the outcome of the experiment. All the patients *think* that they are receiving the medication, although only those in the experimental group actually are. Researchers consider any change in the medical condition of the placebo group to be a direct result of their mental attitude. This change—if there is one—may be due to their feelings about their doctor or the researcher, or it may be due to their belief that they are being given medication that is helping them—or harming them.

Probably in all research the placebo group should be studied as closely as those who get medication, because "placebo" subjects who improve without medication have something to teach us. Their minds or bodies may hold a clue to how we can improve our health without expensive medications that have side effects. This point is relevant not only for studies of ginkgo, but for all medical studies.

Of course, for the mind to have an effect on the body, the brain has to be functioning properly. Certainly it helps to keep the brain healthy and to combat physical and intellectual and emotional deterioration. The Roman writer Juvenal said that a healthy mind in a healthy body should be the goal of everyone. We might say that if the brain is healthy, the mind is healthy. *Ginkgo biloba* is the newly "discovered" ancient herb that makes this possible.

WHAT THE BRAIN NEEDS TO FUNCTION

Proper functioning of the brain depends on a steady supply of glucose (blood sugar) and oxygen, which are brought to the brain—as they are to all tissues—by blood carried through arteries. The nerve cells of the brain require great amounts of glucose and oxygen because they are at work constantly, day and night, whether they are coping with conscious activities, such as driving or cooking, or the unconscious activities of dreams and nightmares. Although the brain accounts for only two percent of body weight, it requires one-fifth of the body's oxygen and has little or no reserves on which to draw.

The Importance of the Blood-Brain Barrier

The brain has a great appetite for nutrients, but it does not need all the substances that circulate in the blood. Some of these can even be harmful to the brain. A filtration system called the blood-brain barrier does not allow "heavy" molecules, such as certain hormones, to come through. Nor does it admit small molecules and ions that have an electric or ionic charge that might impede brain functions. Meanwhile, glucose and oxygen do get through. Thus, the brain is protected from chemicals that it has no need for and that might adversely affect its functioning.

The barrier that regulates the substances that reach the brain from the blood was discovered at the turn of the twentieth century by Paul Erhlich, arguably the greatest chemist of modern times. Erhlich injected dyes into subjects' bloodstreams and found that they traveled everywhere except within the brain and spinal cord. The meninges, three membranes enveloping the brain, were colored with the dyes, but the dyes did not get past this envelope. Erhlich deduced that there was a restriction between the inside and the outside of the brain. It has since been learned that this barrier is created by the capillaries of the brain, which differ from capillaries elsewhere in the body. Cells within brain capillaries deliver a lot of nutrients.

There is also an electrical difference between the inside and the outside of the brain barrier that completely prevents passage of many substances. Substances that are soluble in oil or fat can cross the blood-brain barrier more readily than water-soluble ones. That is why anesthetics that have to reach the brain must be liposoluble, or soluble in fats.

FACTORS THAT PUT THE BRAIN AT RISK

The brain is superior to the computer not only in its versatility and independence, but in its ability to survive. The skull protects the brain fairly well from outside blows, so most injury to the brain comes from within, from the breakdown of the cells or the lack of blood flow. If an area of a blood vessel becomes weak, the vessel can swell like a weak tire, creating a condition called aneurysm, and may burst, resulting in bleeding. If this happens within the brain, the result is a cerebral hemorrhage, which can lead to paralysis or death. Bleeding or a blood clot in a blood vessel supplying the brain can lead to stroke.

Stress

We are subjected to stress that assaults us from every aspect of modern life. It wears us down and distracts us to the point of slowing our mental responses and impairing mental performance.

When we are faced with a stressful situation, anything from making a speech to going through a divorce, the body increases the production of certain hormones to help us cope. Their function is to help us by triggering our "flight or fight" system. This system worked for primitive human beings, who did actually have to protect themselves physically from fearful situations, such as the attack of wild animals, either by confronting them or fleeing from them. However, for the most part, contemporary people cannot fight or flee. We have to look calm, in control, and in charge of ourselves so that we can inspire confidence in our abilities.

Any day can bring problems and sudden crises, ranging from forgetting one's keys to receiving news of the death of a parent over the telephone. One particularly stressful period of time for many people who work in large cities is the time they spend commuting.

Whether you are fighting a crisis or traffic, at some point the rush of hormones brought on by stress confuses your mind and hampers your ability to cope. Despite the primitive "flight or fight" reaction, which evolved as a successful mechanism to help people survive, four-fifths of people are unable to cope with crises such as natural disasters, according to the American Medical Association's *Encyclopedia of Medicine.*

High Blood Pressure

Because the brain requires such a large supply of blood to bring it energy and oxygen, medical conditions affecting the blood and/or circulation can have serious consequences for the brain. As a result of chronic high blood pressure, known as hypertension, the capillaries leak, and plasma, the yellowish blood fluid, seeps out of the capillaries and into the surrounding tissues, which become spongy. If this leakage of plasma occurs in large areas of the brain, pressure can build in the brain and vital fluids can be lost. Dementia, whose symptoms may include memory loss, confusion, emotional outbursts, and embarrassing behavior, including incontinence, results.

Chronic high blood pressure can also reduce the brain's ability to metabolize adrenaline, the hormone that makes the heart work faster to help the body cope with stress, physical exercise, or a strong emotion, such as fear. Without sufficient adrenaline, blood cells may break down and may release their supply of fatty acids, such as arachidonic acid and its metabolites. As we have seen, arachidonic-acid derivatives can generate free radicals and be transformed into substances that create serious alterations in the blood-brain barrier. Everything then goes amok. Capillaries may break, and the person who is this unfortunate circumstance may suffer "mini-strokes."

Lack of Oxygen

In a location where the level of oxygen in the atmosphere is low, such as an airplane cabin or on a high mountain, people can become dangerously lightheaded, or dizzy, because of a fall in the pressure of blood to the brain. The combination of reduced oxygen and reduced atmospheric pressure means that their bodies receive less oxygen, which affects the muscles, heart, lungs, and brain.

Physical Damage

A blow to the head may stun someone, or render a person unconscious, but most shocks to the head cause no permanent damage because the skull protects the brain. However, if the injury is significant, and the brain tissues bleed and/or swell, severe headaches, dizziness, paralysis, and even blindness can result.

WHAT CAN OCCUR WHEN THINGS GO WRONG

A decrease in the oxygen supply harms the brain almost immediately. Such a decrease may be caused by stroke, a blood clot, or hardening of the arteries; it may be a short, severe deprivation or a lesser one of long duration; in every case, the damage is almost immediate. The blood vessels and tissues begin to degenerate, making the brain prey to the formation of tumors and clots. In the brain, ischemia, which is a reduction of blood flow that results in a lack of oxygen, sets off a sequence of harmful changes, including edema (swelling of brain tissues), chemical dysfunction, and the production of free radicals.

When there is a shutdown of blood to the brain cells, there is also a reduction in the supply of nutrients to the cells. The resulting suffering is proportional to how much the cells need the nutrients or energy. This can be compared to what occurs when you do not eat. When you fast, you may not feel well, but you will not die or suffer permanent injury unless you go without food for a very long time. The same is true at the cellular level. The suffering cells send messages that result in increased fluids in the tissues, causing swelling. It is a vicious cycle, because just as the tissues swell in one area of the brain, they are forced to constrict in another because the skull limits how much the brain can expand. In some areas, there is a narrowing of the blood vessels caused by the pressure exerted on them by the swollen tissues.

THE EFFECTS OF AGING ON THE BRAIN

The aging process affects the brain beginning at the age of twenty-five. At about this time, damage to or loss of brain cells begins to make it more difficult to memorize things or learn something new. Reaction time slows, making responses slower.

A decrease in blood flow to the brain as a result of aging leads to a cluster of symptoms commonly associated with older people, including depression; tinnitus, or ringing in the ears; memory loss; and disorientation.

Deterioration of the Blood-Brain Barrier

Over the course of time, as part of the aging process, the walls of the capillaries that form the blood-brain barrier become thicker, thus narrowing the passageway for blood, so that the capillaries bring less oxygen and glucose to the brain, slowing it down and making it less responsive. We have learned this through studies in which scientists have injected radioisotopes into certain blood vessels and then looked at the emission of the isotopes through magnetic resonance imaging (MRI).

Alzheimer's Disease

Alzheimer's may well be the most feared disease, surpassing even cancer in the list of misfortunes we hope to avoid. (See Alzheimer's Disease: A Case History, on page 68, for a description of the human toll of the disease.) It is the cause of

75 percent of cases of dementia in people over the age of sixty-five. It is rare before the age of sixty, but affects more than 30 percent of people over the age of eighty-five.

Many changes occur in the brain in Alzheimer's disease:

- Nerve cells of the brain deteriorate.
- The brain shrinks over a period of time.
- The smaller structures of the brain degenerate.
- Capillaries, those blood vessels that are normally the width of a hair, swell.
- Exchanges between the inside and the outside of the blood-brain barrier are probably affected.

Because research cannot be done on the living human brain, the causes of the symptoms are not known. Theories that are under study suggest that chronic infections, poisonings by toxic substances such as aluminum, and heredity play roles in the disease. About 15 percent of people with Alzheimer's have a family history of the disease.

The one good thing that can be said about Alzheimer's disease is that it has been identified as a disease. This means it can no longer be considered an inevitable consequence of old age and the life cycle.

We have no animal model of Alzheimer's disease. It is one of the diseases that is strictly human. This may be because we have no way to evaluate the memory and behavior of animals. The damage in the human brain created by Alzheimer's disease has no counterpart in animals. There is a sea mollusk called *Aplysia* that acquires memory like mammals, so that is where the experiments are done. Experiments have been done in this creature involving a toxin called neurotoxin af64a, which affects the central nervous system by eliminating a neural messenger called acetylcholine. With neurotoxin af64a, researchers can create something in the lab that is similar to Alzheimer's disease so that they can study the systems.

There is no cure for Alzheimer's disease, so doctors do not

treat the disease, but instead try to make patients function better. Although no one understands its causes or how to cure it, at least the scientific and medical communities are trying to combat it. Until a cure or preventive is found, the best treatment, at least in the early stages, is proper nutrition and rest, and, very possibly, *Ginkgo biloba*. As we shall see, *Ginkgo biloba* has been effective in improving the early stages of the disease when memory loss is obvious, leading to anxiety and depression. Ginkgo is not used in large doses for quick results, but in successive, smaller doses that improve brain functions like memory over a long period of time.

Cerebral Insufficiency

The term *cerebral insufficiency* refers to a group of twelve age-related symptoms that are said by German researchers to be treatable with ginkgo: difficulties in concentration and memory, distraction, confusion, lethargy, fatigue, weakness, depression, anxiety, dizziness, ringing in the ears, and headache. These maladies have been associated with poor blood circulation in the brain, and they are sometimes considered the beginnings of dementia or other degenerative diseases. (*See* Cerebral Insufficiency: A Case Study, on page 72, for an example of the effects of the condition.)

Cerebral Disorders

We have seen that a decrease in the flow of blood to the brain leads to a cluster of symptoms that we associate with older people, including depression, ringing in the ears, memory loss, and confusion. While these conditions are progressive and reach their full negative impact after years of slow deterioration, they can begin subtly in the middle of one's life and then gradually become more obvious.

Memory Loss

How is it that memory begins to go? How can it be that you open a magazine and see a photograph of a large steel vehi-

ALZHEIMER'S DISEASE: A CASE HISTORY

Jeanne walked into her home one afternoon in total perplexity. She had gone downtown an hour before for a specific purpose and just when she was ready to park her car, she realized that she had no idea of what she had intended to do, the purpose that had brought her downtown. She returned home and told her husband that strange story. Later, they learned that she had planned to meet two of her oldest friends for lunch.

Jeanne was then in her mid-fifties. In the following months, she began to forget more and more. She began to write lists of everything she was supposed to do each day, and exactly where she was to go.

In a few months, written reminders were no longer enough. She could not remember the location of the supermarket or her sister's house. She was sometimes confused about her own address when she turned onto her own street. Oddly, she found that memories of the distant past returned to her with great clarity. Although she was unable to recognize one of her neighbors, she could go on at length about events that had happened to her in the fourth grade. Of course, she was depressed, because she knew in her clearer moments that something was wrong with her and that things were only getting worse.

The agony of her husband is easily understood. Not only was he losing the woman he knew and loved, but he was seeing her deteriorate physically and had to learn how to act as her nurse. Then Jeanne began to hallucinate, particularly at night. She saw snakes and trains that were rushing toward her. She lost control of her bowels and wandered aimlessly at night.

In time, her husband needed professional help to take

care of her. A few years after the day when she had returned home from her trip downtown with her strange story about not recalling why she had left their house in the first place, she was confined to her bed. She needed nurses who knew how to cope with bed pans and bedsores and feeding.

No one could be sure that Jeanne's problem was Alzheimer's disease, because that disease can be diagnosed only by a brain biopsy, which is impossible in life; however, all the progressive signs were there: marked forgetfulness, severe memory loss, and severe disorientation.

cle that you have seen on television and in the movies, and even in children's toy shops, but you cannot remember the name for it?

You know that you know the word for it. It is just outside the range of your mind. The word is a simple name that you should be able to recall easily. It is not that the word is terribly important, because this large steel vehicle is not something you have to think of every day. The idea of this thing does not even really matter to you, yet you are sickened by your inability to name it. Suddenly this simple word, this simple label, is not available to you.

You could explain the concept of this vehicle, and you can see it rolling across the screen of your mind as if it were in some old black-and-white film. Maybe your education is such that you can even recall when this large steel vehicle was first developed and why it was important in the history of warfare. Yet your inability to recall the one word that names it is as embarrassing as it is frightening. Something has gone wrong with your memory, and not for the first time. It is one thing to forget this simple term, but what else might you be forgetting?

So you close the magazine with the picture of the large steel vehicle thing, give yourself a few minutes, and finally the phrase "armored tank" comes to mind. Thank heavens! Your brain cells had not died, they were just "sleeping."

There are various explanations for our forgetting what we know, various explanations of why memory fails. These include the following theories:

- Natural organic processes in the nervous system blur memory. (Little evidence has been produced to support this theory.)
- Time itself modifies our memories.
- New learning drives out old knowledge, just as attics must at some point be cleaned out to make room for new storage.
- We repress, or deliberately forget, things that are too painful to remember.

Since memory is such a complex process, it seems most likely that some combination of these factors—and perhaps others not yet identified—is involved in the failure of memory.

Dementia

Many cases of dementia, especially among older people, are caused by Alzheimer's disease, but not all are. In dementia caused by decline of the nerve cells, the decline of mental function is associated with an interruption in the delivery of oxygen or glucose to the brain. This interruption may result in the release of free radicals, with harmful consequences.

PROTECTING THE BRAIN WITH GINKGO BILOBA

Obviously, our ability to make the most of our brains depends upon the brain's being healthy in the first place. The tissues and cells of the brain must be in peak condition, and so must the blood vessels—the arteries, veins, and capillaries that bring blood bearing oxygen and nutrients to our brains. *Ginkgo biloba* is a great aid to this effect.

Preserving the Health of a Healthy Brain

Since we know ginkgo is good for circulation, and since it seems this is the reason it helps to alleviate symptoms of brain disease, it is safe to say it promotes the health of the "normal" brain. When ginkgo is taken as a preventive medicine by people whose brains are relatively young and healthy, it has several beneficial effects that help to forestall the appearance of problems due to aging, including improving mental performance and improving short-term memory.

Improving Short-Term Memory

Ginkgo biloba can improve the mental performance even of relatively young people. In a double-blind study conducted at the Laboratory of Physiology in Paris, eight healthy women ranging in age from twenty-five to forty were given either a placebo or 120, 240, or 600 milligrams of *Ginkgo biloba*. One hour later, they were subjected to a battery of highly sophisticated and reliable tests of short-term memory. The short-term memory of those who had taken 600 milligrams of *Ginkgo biloba* was significantly better than the short-term memory of those who had taken a placebo. Those who took the lower dosages and those who received the placebo were statistically about the same in test results, according to the report in *La Presse Médicale* of September 25, 1986.

Ginkgo biloba definitely improves short-term memory when taken in high dosages. Significantly, in all these volunteers there was not a single side effect. Younger people who do not have much confidence in their short-term memories should take one high dose of *Ginkgo biloba* before an exam, such as a driver's license test. There will be no side effects, the ginkgo will help relieve the person's anxiety, and, judging from the results of objective tests, short-term memory should be better.

Tips for Learning Well and Remembering

Much of what we forget, be it geometry or what we meant to buy at the store, we forget because we did not pay attention

CEREBRAL INSUFFICIENCY: A CASE STUDY

Joseph was a man who enjoyed life. He had a job in sales that he enjoyed and did well. As he matured, he moved into management and became a key executive for a major insurance firm. Basically, he thought that taking care of one's health was for sissies, and he ate and drank whatever he cared to. The only positive things he did for his health were playing tennis avidly and avoiding smoking.

As an athlete, he was more enthusiastic than talented. Nonetheless, he played at least once a week, usually with players at his own level, and sometimes with better players who enjoyed their games with Joseph because he had an interesting and sparkling personality.

Even after he retired from his work—somewhat against his will—he and his wife continued to lead an active social life. They traveled the world, enjoyed their grandchildren, and enjoyed moving to a resort area after they sold their home in the snowbound Northeast. Joseph's physical health was good, and despite his increasing years, everything about his life was enjoyable—he even found volunteer work that called upon his executive abilities, widened his circle of friends, and won him several community awards.

However, for the first time in his life, around the age of 72, he became horribly depressed. Joseph began to experience painful cramping in his legs that made it difficult for him to walk, much less play tennis. He saw no point in continuing to do volunteer work and decided to stay at home during the day rather than encounter anyone he knew for whom he would have to pretend to be his old self.

For a while, having a puppy helped. Joseph enjoyed the puppy's antics, and its affection, and the simple acts of taking care of the dog—feeding it and taking it for walks—lifted his mood, even though the pains in his legs prevented Joseph from walking far. Eventually, though, the novelty of the puppy's tricks wore off, and it grew from a puppy to a dog. Soon Joseph's mood was worse than ever.

His children thought Joseph was having a problem coping with getting older and facing the fact that most of his happy life was behind him. He absolutely ruled out any kind of psychotherapy. When his wife finally convinced him to go to a general practitioner, tests showed that Joseph had serious hardening of the arteries. That was the cause of his leg pains, but more amazingly, it was also the source of his mood problems. His brain was not getting enough oxygen and nutrients, and as a consequence it was not performing as well as it should have been. With the areas of his brain that controlled his mood starved for nutrients, his emotional health was as damaged as the nerves of his legs.

Opening the arteries of his brain helped to improve his mood. He did not become any wiser or more philosophical, but he was able to function appropriately. To accomplish this improvement, Joseph had not had to confront issues of his childhood that had existed sixty-five years before; he simply needed to make sure his brain got the blood supply it needed. *Ginkgo biloba* helped to solve those problems and return him to his upbeat outlook on life. If he had taken it earlier, it could probably have helped him avoid depression in the first place.

to it in the first place. Taking time to pay attention to the people who are introduced to us, noting the features of their faces, and making some associations to help us link their names to their faces result in a better memory for names.

Although it is true that mental ability decreases somewhat

naturally after the age of twenty-five, reviewing and deliberately recalling information offsets this natural decline. Studies have demonstrated that eighty-year-olds learn a language as easily as fifteen-year-olds if people with poor brain circulation and other diseases are excluded from the study!

During learning, the mind needs periods of activity and rest. If the right balance between them is achieved, recall improves. If you study for a bit, then pause to let information "sink in," you will have a better ability to recall what you have read or studied. Learning periods should consist of twenty to forty minutes of reading, including note-taking. Then, a ten-minute rest should follow, succeeded by a ten-minute period to recall the information, with a check of the original notes to make sure that you have recalled the data correctly. Your memory will be reinforced if you review important information for a few minutes each day. Details of the material will be lost unless they are reviewed.

Of course, both learning and remembering depend on the healthy functioning of the brain, and *Ginkgo biloba* can help keep the brain healthy.

Strengthening Brain Cells

We know that in humans, *Ginkgo biloba* protects the tissues of the brain from harm. Ginkgo prevents the breakdown of the cells by strengthening their membranes and by scavenging free radicals. Studies show that *Ginkgo biloba* activates the enzyme that exchanges salt for potassium in the red blood cells. This exchange improves the energy or electrical charge of the cell membrane, which is particularly important in nerve cells and the central nervous system.

Scavenging free radicals is accomplished by the antioxidant action of ginkgo. Free radicals are produced naturally in the body, and under the best of conditions they are beneficial, as when they help to fight off infection caused by invading microbes. However, when free radicals are generated too quickly, they set off a chain reaction of unhealthful biochemical processes.

Ginkgo's ability to scavenge free radicals is most obvious in the brain cells because they contain the highest percentage of fats. Remember that the blood-brain barrier admits certain substances to the inner area of the brain only if they are soluble in fat. The brain contains the highest percentage of fatty acids of any organ in the body. Fatty acids give the brain its vitality because they have a high level of energy, but they also make brain cells vulnerable to attack from free radicals, which are considered the major cause of aging of the brain. Free radicals are drawn to fatty acids, especially when the supply of oxygen to the brain is slowed. When free radicals react with fatty acids, they form lipid peroxides that damage the cells.

Increasing Blood Flow to the Brain

In the last fifteen years, increasing research into the effects of *Ginkgo biloba* has improved our understanding of the herb's astonishing ability to improve blood supply to the brain, thereby sharpening mental function and, equally importantly, protecting the brain from deterioration. We know that *Ginkgo biloba* allows the tissues of the brain to get enough blood, oxygen, and energy.

Taking *Ginkgo biloba* improves the ability of the blood vessels to deliver glucose because it keeps the vessels functioning properly. In the next chapter, we will go into greater detail on how *Ginkgo biloba* helps the blood vessels of the entire body, but for now we shall concentrate on its impact on the brain.

Ginkgo biloba regulates free radicals; that is, it prevents dangerous free-radical formation in the brain and circulatory system and improves the flow of the blood by maintaining the blood vessels. *Ginkgo biloba* is the greatest boon to blood vessels that we know of. It does several things for blood vessels. It maintains their strength by helping to deliver glucose to the right targets in the cells of the blood vessel walls. It also cleans up free radicals and neutralizes harmful substances in the blood.

"No other known circulatory stimulant, natural or synthetic, has selectively increased blood flow to disease-damaged brain areas," according to *Alternative Medicine: A Report to the National Institutes of Health on Alternative Medical Systems and Practices in the United States.* That report noted that in Europe, ginkgo is used mainly to counteract the symptoms of aging. The report continues, "It is believed to stimulate circulation and oxygen flow to the brain, which can improve problem solving and memory. It was shown to increase the brain's tolerance for oxygen deficiency and to increase blood flow in patients with cerebrovascular disease."

Ginkgo biloba balances the actions of the arteries and veins, relaxing them in times of spasm and stimulating them in times of paralysis. This dual action is particularly crucial in the brain because so-called "single direction" drugs, which have only one action, can harm healthy areas of the brain. For example, a drug that opens, or dilates, the blood vessels will open all the blood vessels, even healthy ones. Thus these drugs can have a bad effect on healthy tissues by allowing too much blood to flow to them, drawing blood and oxygen away from the damaged, oxygen-deprived tissues that the drugs are supposed to be helping.

Dr. Eric Lasserre, who works in the French Alps, was the chief physician for French expeditions to the Himalayas, the highest mountain range in the world. A few years ago, when some people wanted to climb the range without taking oxygen bottles, Dr. Lasserre recommended that they take *Ginkgo biloba* extract. The climbers found that they could climb higher and for longer periods of time without experiencing "mountain sickness."

Minimizing the Consequences of Damage

Ginkgo's anti-free-radical properties not only prevent damage to cell membranes but restore damaged cell membranes throughout the body. Recent experiments with laboratory animals indicate that ginkgo may actually help to regenerate damaged nerve cells. Researchers in Marseilles reported in the journal *Pharmacology, Biochemistry & Behavior* in 1991 that cats with damage to their central nervous systems showed signs of recovering when given *Ginkgo biloba* for thirty days. Their balance, posture, and ability to move their necks improved more rapidly than these functions improved in similar cats that were not given the extract.

Ginkgo biloba may be effective in treating injuries to the spinal cord. In a study published in the German journal *Recent Results in Pharmacology* in 1995, Turkish scientists from the School of Medicine in Kayseri treated one group of paraplegic rats with a combination of a corticosteroid drug, a thyroid-releasing hormone, and *Ginkgo biloba* extract. They left a control group untreated. They found that the rats taking *Ginkgo biloba* and the steroid methylprednisolone showed greater improvement than the untreated group. The scientists noted: "These results suggest that methylprednisolone and *Ginkgo biloba* may have a protective effect against ischaemic spinal cord injury by the antioxidant effect."

Obviously, *Ginkgo biloba* would be a good treatment for those who have suffered an injury to the brain or head. Better yet, it makes a good preventive, keeping the brain and all its blood vessels able to resist the kind of trauma that can occur in an automobile accident, even a minor one, or on a football field. Some of us, particularly athletes and construction workers, are more at risk for damage than others, but all of us would do well to take *Ginkgo biloba* to help us to resist damage to our all-important brains.

In 1986, French researchers reported their results on tests in this area in *La Presse Médicale*. They caused the brains of lab animals to swell and then gave one group of the subjects an oral dose of *Ginkgo biloba* and the other group a dose of water. They found that eighteen hours later the swelling had disappeared in the animals treated with ginkgo, whereas those treated with water showed no change. In another experiment, cited in the same issue of *La Presse Médicale*, two groups of rats were given poisonous tin salts in a solution of water. One group was also given *Ginkgo biloba* extract. After two weeks, the brains of the rats in the first group swelled, and their nervous systems showed obvious signs of damage. They could not climb, and they walked on their toes instead of their feet.

The tin salts had affected the membranes of their brain cells and allowed fluids to leak out of cells and swell the tissues. However, the rats that had been given *Ginkgo biloba* extract along with the tin salts showed no change in behavior. Ginkgo kept the cell membranes intact. It prevented leakage. If given ginkgo early enough, the cells resist the damage, first because of the protective effect of *Ginkgo biloba* on cell membranes in general, especially those of the brain; and second because ginkgo protects normal factors of the blood vessels by making them able to resist swelling (and narrowing—as we shall discuss in the next chapter).

Researchers at Rutgers University in New Jersey reported in the journal *Experimental Neurology* in 1989 that after thirty days of treatment with *Ginkgo biloba* extract, swelling was reduced in rats that had suffered injury to the brain. People who have suffered concussions from sports injuries or other accidents to the head would do well to take *Ginkgo biloba* until their symptoms disappear. By making the cells strong enough to withstand oxygen deprivation, and increasing blood flow to the brain, *Ginkgo biloba* works to prevent or minimize damage to the brain.

Combatting Cerebral Disorders Due To Aging

Ginkgo biloba extract can increase the rate at which information is transmitted on the nerve-cell level. Studies conducted on laboratory animals at the Faculty of Medicine and Pharmacy in Poitiers, France, have shown some promising results in this regard. One found that 100 milligrams of ginkgo taken daily for four weeks increased the number of receptors in the brains of aging (two-year-old) rats. The results were reported in the *Journal of Pharmaceuticals and Pharmacology* in April 1994. Another French study compared elderly rats (twenty-four months old) with young rats (four months old) and reported in 1994 that *Ginkgo biloba* reversed the clinical signs of aging in the brains of the animals and revitalized them.

Perhaps most dramatic was a 1986 French study led by J. Tallendier and reported in the September 25, 1986, issue of *La*

Presse Médicale, concerning 166 patients over the age of sixty who had lived in retirement homes for at least a month. Researchers evaluated the subjects in seventeen key areas:

- Level of liveliness or vivacity.
- Degree of anxiety or lack of worry.
- Degree of depression or elation.
- Emotional stability or instability.
- Keenness of short-term memory.
- Presence or absence of vertigo.
- Frequency and severity of headache.
- Presence or absence of ringing in the ear.
- Ability to orient oneself in relation to surroundings.
- Willingness to take initiative.
- Level of cooperation with others.
- Level of social interaction or sociability.
- Maintenance of personal hygiene.
- Degree of ability to walk.
- Appetite.
- Frequency of fatigue.
- Sleep patterns.

The subjects had obvious symptoms, but were not severely damaged or too far gone into dementia to be able to comment on their reactions and help with evaluations. They were tested at three, six, nine, and twelve months, during which time one group took 160 milligrams of *Ginkgo biloba* daily and another group took a placebo. After three months of study, those taking *Ginkgo biloba* extract showed a marked improvement that continued and increased as time passed.

After twelve months, those who had been given *Ginkgo biloba* were almost normal. Those who took the placebo improved for a short time, and then hit a plateau. The amazing thing was that among those taking ginkgo, patients who were the worst off and had suffered from problems for the longest periods of time showed the most improvement. It seems paradoxical, but where there is life there is hope.

Anyone who has problems with alertness, short-term memory, headaches, balance, mood, or tinnitus should consider taking *Ginkgo biloba*, even if the conditions have been deteriorating for years. We all owe it to ourselves to follow the research on ginkgo and how it combats the effects of aging on the brain. New and exciting developments are on the horizon.

Combatting Alzheimer's Disease

At this time there is not sufficient evidence to say that *Ginkgo biloba* extract is useful in the treatment of Alzheimer's disease. However, it does seem to slow the progression of symptoms if taken in the early stages. *Ginkgo biloba* extract may delay the onset of these conditions so that those susceptible to Alzheimer's can to continue to live normally, without being a burden to loved ones and thus escaping institutionalization.

Alzheimer's disease is not just a disease of the brain, but of the entire system, even extending to the bone marrow. Studies show that people who take ginkgo for one or two months have an increase in the turnover of the neurotransmitter norepinephrine in the brain, indicating increased activity in the brain. However, ginkgo does not affect serotonin, another neurotransmitter, which is involved in regulating consciousness and mood, digestion, and even the clotting of the blood.

In a month-long German study published in 1996, 216 patients suffering from mild to moderate symptoms of Alzheimer's were divided into two study groups. The subjects in one group were treated with 240 milligrams of *Ginkgo biloba* extract daily, and the subjects in the other group were given a placebo. At the end of the study period, the subjects were tested for mental, behavioral, and motor skills. Those who had taken ginkgo showed a great increase in mental alertness and improvement in mood, so the effectiveness of the *Ginkgo biloba* extract in treating the symptoms was confirmed, as reported in *Pharmacopsychiatry* in March of 1996.

Improving Memory

When we discussed the role of ginkgo in preserving the health of a healthy brain, we said that a healthy person's memory problems usually can be corrected if certain learning techniques are employed to make the mind work properly. Moving beyond those techniques, European researchers have probed the scientific reasons for the traditional use of *Ginkgo biloba* to stimulate memory.

The Laboratoire de Pharmacologie Clinique in Rennes, France, studied eighteen men and women around the age of seventy who were experiencing slight difficulties with their memories. In this study, the same group of subjects was tested several times, with different doses of *Ginkgo biloba* extract and also with a placebo. When the subjects were tested one hour after taking 320- or 600-milligram doses of *Ginkgo biloba*, they were significantly more alert and quick-witted than they were when they had swallowed only the placebo. They were able to respond twice as quickly as they did without ginkgo. They showed a significant increase in the speed with which they processed information, even at the age of seventy, according to the report, which was published in the May-June 1993 issue of *Clinical Therapy*.

English researchers studied the effects of *Ginkgo biloba* on thirty-one patients over the age of fifty who showed mild memory loss. Those who received 40 milligrams of the extract three times per day were tested at the end of twelve and twenty-four months and were shown to have improved mental function and quicker response time than those who did not take the extract. This was reported in *Current Medical Research and Opinion* in 1991.

Researchers at three different test centers in Germany studied the effects of *Ginkgo biloba* on seventy-two people who were showing loss of memory and mental function. They found a statistically significant improvement in short-term memory after six weeks of treatment and improvement in the rate of learning after twenty-four weeks. Those who took a placebo rather than *Ginkgo biloba* showed no improvement, as reported in *Fortschritte der Medizin*, February 20, 1992.

Bulgarian scientists discovered that *Ginkgo biloba* extract improved the retention of learned behavior, which involves both learning and memory. As the Bulgarians reported in *Planta Medica* in April 1993, they fed aged rats (twenty-six months old) and young rats (three months old) doses as high as 240 milligrams per kilogram of weight daily. The degree of progress varied with the dosage and quality of the extract.

The results of a German study reported in *La Presse Médicale* echoed those of the French short-term memory study cited earlier (see page 71). Eight healthy women between the ages of twenty-five and forty were given *Ginkgo biloba* extract in doses of 120 milligrams, 240 milligrams, or 600 milligrams. The tests were so reliable and the results so obvious that this small group makes for a significant study. They were given a battery of tests to measure their short-term memory. One hour after taking the extract, the short-term memory of those who took 600 milligrams of *Ginkgo biloba* showed significant improvement over the short-term memory of those who had taken the placebo. The effects from the lower dosages and the placebo were about the same. It appears that short-term memory is definitely improved by ginkgo if the dose is high. In all these volunteers, there was not a single side effect.

British scientists at the University of Leeds have also added to the medical knowledge of *Ginkgo biloba*. Over a six-month period, they studied the effects of *Ginkgo biloba* on thirty-one patients over the age of fifty who showed mild memory loss. Those who received 40 milligrams of ginkgo extract three times per day were tested at the end of twelve and twenty-four months and were found to have improved mental function. They also had quicker response times than those who did not take the extract, according to the report of this study in the September 26, 1986, issue of *La Presse Médicale*.

These researchers also studied other drugs. They found that amphetamines did not improve memory and that sedatives actually destroyed short-term memory. Some anti-depressants seemed to improve short-term memory, but the researchers were unable to determine why.

Combatting Dementia

We know that ginkgo is useful for patients displaying the first symptoms of dementia. Cerebral aging is different from dementia, but the symptoms are equally worrying. The causes of all these problems are unknown. At this point, it is useful to develop research that seeks suitable treatment and to base this research mainly on preventive therapy—on ways to keep the brain tissues, the blood, and the blood vessels that carry it healthy from early adulthood on to old age, so that chronic, incurable problems will not begin.

In yet another German study, twenty patients between the ages of sixty and eighty-seven who were experiencing various symptoms of dementia were given 120 milligrams of *Ginkgo biloba* extract per day. Unlike a similar group that had been given a placebo and showed no change at all, the group given *Ginkgo biloba* improved dramatically in tests of various clinical geriatric scales.

Another group of researchers studied patients with dementia who were still living at home. They were either presenile or senile due to Alzheimer's or because of blockage in the flow of blood to many areas of the brain. The study was double-blind, multi-centered, and included a placebo group, so it was the most rigorous kind of test we can set up. The researchers started with 262 subjects and observed them for four weeks. The subjects were divided into two groups. One group received a daily dose of 240 milligrams of *Ginkgo biloba*, the other a placebo. The researchers then conducted a series of tests to evaluate the subjects' mental function. These included the Clinical Global Impression for psychopathological assessment; the Syndrom-Kurztest (SKT) for assessment of attention and memory; and the Nurnberger Alters-

Beobachtungsskala to determine the activities of daily life. By these measurements, the researchers could make the most precise determination of how the subjects were really doing. In this series of tests, patients had to improve in two of three areas of mental functioning to register a positive response. In 156 patients who completed the test, the rate of improvement of those who took *Ginkgo biloba* was significantly better than that of the subjects who took a placebo. These findings, reported in the March 1996 issue of *Pharmacopsychiatry,* are particularly significant considering the good scientific model and good statistical methods used.

Ginkgo also worked for one ninety-six-year-old woman who suffered from dementia. She was given 120 milligrams of *Ginkgo biloba* daily. After three weeks, six weeks, and twelve weeks she was tested with an electroencephalogram (EEG). An EEG is the best, most objective way to study the health of the brain. An electroencephalogram records the electrical activity of the brain, whether the subject is a person or an animal, awake or asleep, rational or confused. The woman's alertness slowly but clearly improved as she took *Ginkgo biloba* extract, as was shown on her EEG. This study was conducted at the Laboratory of Physiology in Paris and reported in the September 26, 1986, issue of *La Presse Médicale.*

The first U.S. study on the use of Ginkgo biloba extract for the treatment of Alzheimer's- and stroke-related dementia, reported in the October 22, 1997, issue of the *Journal of the American Medical Association,* also showed promising results. In this study, undertaken at the New York Institute for Medical Research in Tarrytown, New York, subjects with dementia were divided into two groups. One group was treated with 120 milligrams of a European *Ginkgo biloba* extract daily; the other was given a placebo. Before treatment began, the subjects were tested by means of the Alzheimer's Disease Assessment Scale, a standard test that measures cognitive function. The test was then repeated after twelve weeks, six months, and one year of treatement. Alzheimer's patients normally show an average decline of four points per year on this test, but among the subjects in this study, 27 percent of those taking ginkgo registered a four-point *improvement* after one year, compared with a fourteen-point improvement among the placebo group. The subjects' functioning was also assessed by their caregivers, who reported improvement in 37 percent of the ginkgo patients, versus 23 percent in the placebo group. After the conclusion of the study, the subjects who had been given a placebo were given the opportunity to take ginkgo extract as well.

Combatting Depression

The German Ministry of Health Committee for Herbal Remedies has approved the use of ginkgo extract as an herb traditionally used for improving mood and mental processes, and German biomedical researchers have investigated its effectiveness in this area. One study, reported in the German journal *Arzneimmittel-Forschung* in 1985, found that elderly individuals who did not respond to anti-depressants showed marked improvement when they were given 240 milligrams of *Ginkgo biloba* daily. After four weeks of treatment they were significantly happier, more optimistic, and more motivated. Improvement continued during the eight weeks of the trial study.

Combatting Cerebral Insufficiency

The major claim for *Ginkgo biloba* is that it helps to counteract "cerebral insufficiency" (see page 67). In 1990, German scientists reported that they had treated sixty hospitalized patients who were suffering from cerebral insufficiency and depression with *Ginkgo biloba*. For six weeks they gave the patients a daily dose of 160 milligrams. The patients were tested after two, four, and six weeks of treatment. Those who had been given a placebo showed small but progressive improvements. Those who were given the ginkgo extract showed markedly better results. After two weeks, some of their symptoms disappeared. Then, in the next two weeks, improvements became noticeable. After four and six weeks, eleven of twelve symptoms of patients taking *Ginkgo biloba* had improved, according to the report of the study published in *Fortschritte der Medizin* in 1990.

In reviewing the various studies on *Ginkgo biloba* extract, the distinguished British medical journal *Lancet* reported in 1992, "We conclude that ginkgo extract can be given to patients with mild to moderate symptoms of cerebral insufficiency."

SUMMARY

In this chapter, we have considered how the brain functions. We have looked at its miraculous construction, which includes the blood-brain barrier that filters out substances that benefit the rest of the body but might help the brain.

In the normal, healthy brain, *Ginkgo biloba* improves mental performance. It enhances memory and all brain functions, including alertness and mood, because it keeps the blood vessels healthy and enhances blood supply to the brain.

Time, or the aging process, begins to take a toll on the brain when we are twenty-five, but by using proper learning techniques and by keeping the brain healthy, especially with *Ginkgo biloba*, we can offset this process. Alzheimer's disease, which is possibly the most dreaded malady in the modern world today, is incurable, yet people in its early stages can be greatly helped by taking *Ginkgo biloba*.

Ginkgo stabilizes cell membranes. It scavenges free radicals. It inhibits edema due to damage to the cells of the nervous system. If ginkgo is taken early, it is the only treatment we know of that works for cerebral insufficiency.

If mental function—alertness, attention, memory, and so on—is slowed because of poor circulation to the brain, or because a person is depressed, *Ginkgo biloba* is an effective treatment. In the next chapter, we will consider how ginkgo improves circulation to all the other areas of the body, and thus has an astonishing variety of positive effects on many different aspects of health and well-being.

CHAPTER 5

Ginkgo, the Heart, and Circulation

If there were an intergalactic award for best design, it would surely go to the human body. The collection of molecules, cells, and systems that makes up each person is a brilliant piece of organization and interaction. The brain is its most amazing component, but the heart and circulatory system come in a close second.

In the last fifteen years, interest in *Ginkgo biloba* has increased primarily because the Western world learned that the herb has an astonishing ability to increase circulation. The blood vessels make it possible for the blood to both feed and cleanse the entire body. *Ginkgo biloba* makes the circulatory system more efficient by improving the tone and elasticity of those blood vessels—the arteries, veins, and capillaries. The effectiveness of ginkgo in increasing circulation is such that one 1989 German study found a 57-percent increase in blood flow under the fingernails (the area called the nail-fold capillaries) one hour after subjects took ginkgo.

Adequate blood flow to the muscles ensures adequate oxygen for the cells of the muscles. Without adequate oxygen, the muscles may contract in spasms—sudden, involuntary convulsive movements. Spasms that are especially strong and painful are called cramps. Because *Ginkgo biloba* increases the flow of blood to the arteries of the legs, it is not surprising that painful cramping in the legs lessens, or goes away entirely, after a person takes ginkgo.

Further, ginkgo is valuable because it is a "blood thinner" that inhibits abnormal blood clotting. In this way, ginkgo helps prevent stroke and heart attacks, among other maladies.

A BRIEF TOUR OF THE CIRCULATORY SYSTEM

Ginkgo has been shown to improve circulation, but before we go into the medical findings, let us understand how the circulatory system works. The design for the circulatory system is remarkable. The arteries carry oxygen and nutrients to all the cells. The hair's-width capillaries enable the blood to reach deep into the tissues, where these nourishing substances are deposited, and then to pick up cellular waste products for removal. The veins then carry the blood off to the lungs, where it is replenished with oxygen. If we could isolate a particular portion of the blood and follow it through the circulatory system, we would find that it can complete this journey through the body in as little as thirty seconds.

The circulatory system consists of the following basic components:

- The heart.
- The blood vessels (arteries, veins, and capillaries).
- The lymphatic system, which joins the fluid bathing the tissue cells to the bloodstream.

The circulatory system distributes blood throughout the body so that it can perform its essential functions:

- Bringing oxygen from the lungs to the cells and removing carbon dioxide.
- Bringing nutrients absorbed from the intestine and taking waste materials to the liver and kidneys to be broken down and excreted.
- Bringing hormones and other body chemicals from various glands to every part of the body.

The heart is the engine that keeps this system going. The heart is a pump with four containers, or chambers. These are the right and left atria (from the Latin *atrium*, meaning "corridor") and the right and left ventricles (from the Latin *ventriculus*, meaning "a little belly"). In a single heartbeat, the right side of the heart pumps the "used" deoxygenated blood from the tissues of the body to the lungs, where the red blood cells release carbon dioxide and pick up more oxygen. Within the same beat, the left side of the heart pumps the "fresh" oxygen-rich blood from the lungs out to organs and tissues where the red blood cells release the oxygen and pick up carbon dioxide. The force generated by the heart is very great. If the body's largest artery, the aorta, were to be cut, a six-foot (two-meter) jet of blood would stream forth.

The arteries are the blood vessels that carry oxygenated blood to the body. These are thick and muscular, and lie deep in the body where they are most protected. Only at the wrists are there arteries close to the surface of the skin. The arteries branch out in finer and finer sizes to carry blood to all the tissues. The smallest blood vessels, reaching deep into tissues, are called capillaries. They connect the smallest arteries with the smallest veins. Within the tiny, thread-like capillaries, the transfer of oxygen and nutrients for carbon dioxide and waste products takes place. The veins carry the "used blood" back to the heart, where it is pumped to the lungs for a fresh dose of oxygen.

The circulatory cycle continues without interruption every day of our lives. The health of the blood vessels is crucial to top performance of the human body because every organ depends on oxygen and nutrients to survive and to perform its essential functions. If the circulatory system is not functioning properly, crucial areas of the body may suffer. For example, a decrease in the supply of blood to the heart can cause chest pain or a heart attack.

Because the arteries, capillaries, and veins work so closely together, any malfunction of one component disrupts the whole system. Arteriosclerosis, or hardening of the arteries, is a disease of the walls of the arteries that has been regarded as a consequence of growing old. Like muscles, arteries are supposed to be strong and flexible. Their inside lining should be as smooth as the silkiest skin, so that blood flows freely. Arteries affected by arteriosclerosis, however, are stiff and inflexible, and often have irregular deposits of sticky plaque on their interior walls. Arteriosclerosis is aggravated by high blood pressure.

Atherosclerosis is related to arteriosclerosis, but is an even more serious disorder of the circulatory system. In this condition, fats and cholesterol accumulate in the arteries, narrowing the passageways available to the circulating blood. At the restricted passages, particles in the blood may clot together, restricting the flow of blood further or even blocking it completely. If atherosclerosis occurs in the legs, walking becomes difficult or impossible because the muscles of the legs cannot get enough oxygen to function properly. The result is pain and weakness in the legs. At any location in the body, narrowed blood vessels may rupture, causing a hemorrhage that requires emergency treatment.

Ginkgo biloba benefits the blood vessels by promoting the health of the cells of the blood vessel walls, as we shall see in the rest of this chapter.

THE HEART

The heart is at the center of the human body. It is the most important organ of the circulatory system, and its health promotes the vitality of all its vessels.

Coronary artery disease (disease of the arteries that supply the heart) is the main cause of death in the United States. One of every two deaths in this country results from heart disease, and approximately 6 million people suffer from some of its symptoms. Particularly since World War II, researchers have worked to understand how to protect the heart, and they have made great strides. We now know two crucial things. First, a proper diet, low in fats, maintains the health of the heart. Second, smoking is one of the worst things you can do to this vital organ. We also know that high blood pressure stresses the heart and must be controlled. These factors in coronary health are particularly important for those who have a family history of heart disease.

Surprisingly, living with stress or pressure does not seem to harm the heart, although it makes life more difficult. The old idea that personality type—so-called Type A and Type B—affects susceptibility to heart attacks has been all but dismissed.

Happily, the incidence of coronary artery disease and deaths from heart attacks have decreased significantly since the mid-1960s, due largely to our increased knowledge of the dangers of smoking and of fats in the diet. We are also aware that staying trim and exercising promote the health of the heart. In general, people have applied this knowledge to make healthful adjustments in their lives. Overall, Americans use less tobacco, make better use of medical care, get more exercise, and eat less fat than they did twenty years ago. To all this we can add another weapon in the arsenal against heart disease—*Ginkgo biloba* extract.

A 1992 German study reported in the *Arzneimittel-Forschung* in April of 1992 provided evidence that *Ginkgo biloba* extract, taken over a long period, can reduce cardiovascular problems, including heart disease, high blood pressure, excess cholesterol, and diabetes.

In 1995, a Chinese study reported that the free-radical-scavenging properties of *Ginkgo biloba* may be helpful to the heart. The coronary arteries of rabbits suffering from oxygen deprivation were injected with ginkgo extract. Compared with rabbits injected with a saline solution, those treated with *Ginkgo biloba* were less severely affected by ther heart problems. The heart tissues of those given ginkgo were less swollen and torn than the heart tissues of those that did not receive the extract. These results, reported in *Biochemistry and Biology International* in January of 1995, suggest that the antioxidant properties of *Ginkgo biloba* protect the heart, and ginkgo's ability to help the blood heal damaged vessels helps the heart resist damage.

The Department of Molecular and Cell Biology at the University of California–Berkeley reported similar findings in *Free Radical Biology and Medicine* in 1994. Scientists there found that *Ginkgo biloba* helped preserve the middle layer of the wall of the heart after it was deprived of oxygen.

THE ELEMENTS OF THE BLOOD

Blood not only carries vital supplies throughout our bodies, but plays a major role in the body's defense against infection. The blood is a powerful warrior that has the power, through clotting, to seal damaged vessels that would allow the loss of blood from the body. Platelets that come into contact with the injured site are stimulated, become sticky, and stick together. Thus, the blood makes emergency repairs as well as providing the nutrients needed for the healing process.

Every adult has about ten pints (five liters) of blood in his or her body. This vital fluid is made up of various elements, each of which makes a unique contribution to our well-being. Nearly 50 percent of the blood is composed of red blood cells, or erythrocytes, which develop in the marrow of our bones. A red blood cell is shaped like a doughnut without a hole, but with a depression in the center where the doughnut's hole would be. Red blood cells carry oxygen from the lungs to the tissues throughout the body. Their shape aids this process in two ways: It gives them a large surface area that helps them to absorb and release oxygen molecules, and it makes them flexible enough to squeeze through the tiny capillaries. Variations in the structure of red blood cells form the basis for establishing various blood types (A, B, AB, and O).

Each red blood cell contains hemoglobin, a substance that gives the cell its red color and enables the cell to carry oxygen. Hemoglobin contains iron, enzymes, and nutrients that the cell needs to maintain itself.

The process of forming a red blood cell in the bone marrow takes about five days and requires a good supply of iron, folic acid, amino acids, vitamin B_{12}, and erythropoietin, a hormone produced in the kidneys. After a newly formed red blood cell is released from the bone marrow, it requires two more days to mature before it is ready to perform its function as a carrier of oxygen and carbon dioxide. An individual red blood cell circulates for about four months. As the cells age, they wear out, becoming inflexible and eventually disintegrating. The fragments of disintegrated red cells are trapped in the spleen and other organs where some scavenger cells destroy them. If this entire process, from red-cell formation in the bone marrow through red-cell destruction by the scavenger cells, does not proceed normally, anemia (a deficiency in the blood's oxygen-carrying capacity) can result.

White blood cells, or leukocytes, fight infection, working first to prevent it and then to overcome it if it occurs. They are twice as large as red blood cells, but spend less time in the blood. The three main types of white blood cells are granulocytes, which are formed in the bone marrow; monocytes, formed from the cells lining the capillaries in some organs; and lymphocytes, formed in the lymph nodes. Granulocytes destroy bacteria, and monocytes and lymphocytes participate in the immune response. Some of the lymphocytes are responsible for the hypersensitivity to foreign substances that may result in allergy.

Platelets are solid elements in the blood. These tiny bodies survive for about nine days. Platelets travel in the body in an inert state, but if a vessel is injured, platelets stick to the vessel walls at the site of the injury and to each other, forming a clot that stops the bleeding. The platelets' ability to form clots is welcome when the clotting stops bleeding, but, unfortunately, clots can also be dangerous. If fatty deposits accumulate in the arteries, platelets in the blood may form clots at the site of the deposits. These clots are microscopic at first, but they may grow until they slow the flow of life-giving blood, and they may block an artery at some point. Such a blockage is a condition called embolism.

Plasma accounts for most of the volume of the blood. It is a yellowish fluid that consists mostly of water and carries proteins, fats, glucose, and salts, and, of course, blood cells and platelets, throughout the body.

GINKGO'S EFFECT ON THE BLOOD

The plasma is the first component of the blood to be affected by *Ginkgo biloba*. Research indicates that *Ginkgo biloba* enters the body in the duodenum, the beginning of the intestinal tract. Within three hours of its absorption, it is fully present and working in the plasma.

The anticoagulant effects of *Ginkgo biloba* are derived from its ginkgo flavone glycosides (see Chapter 3). These compounds improve circulation and prevent blood clots by reducing the "stickiness" of the platelets in the blood—that is, their tendency to coagulate, or clot.

Ginkgo biloba has an important positive effect on the endothelium, the layer of cells that lines the heart, blood vessels, and lymphatic ducts. The cells in the endothelium are flat and thin, like stones in a river bed, and furnish a smooth surface that allows blood to flow and helps prevent the formation of clots.

Another of ginkgo's positive effects on the blood is derived from the fact that it helps the adrenal glands perform their important regulatory functions. This pair of glands secretes hormones directly into the bloodstream. Hormones produced by the outer region of the adrenal cortex regulate the body's metabolism, blood composition, and even body shape. The inner region has a separate function. It produces hormones that are the body's first line of defense against stress, whether it be physical or emotional. This inner region of the adrenals is called the *adrenal medulla* and is considered to be part of the sympathetic nervous system. Ginkgo promotes the work of the adrenal glands because it helps the cell receptors that absorb adrenal hormones. In this way it aids the adrenals in regulating metabolism, the chemicals in the blood, and even the shape of the body.

Thus we can see that, even though I have discussed the importance of ginkgo to circulation separately from its value to the brain and nervous system, in fact these systems are interconnected. The Chinese have long considered the body to be one entire system that should be treated as a unity. A rash, for example, indicates to a Chinese herbal doctor that the body is having trouble functioning. The Chinese doctor will try to balance the entire system to eliminate the underlying cause of the rash. A Western doctor, however, is more likely to see a rash as an allergy, a malfunction of the skin or nervous system, and will focus treatment only on the affected area of the body.

FREE RADICALS IN THE BLOOD

Free radicals occur in body tissues as a natural result of normal biochemical processes. Normally, enzymes and other natural substances in the body act as free-radical scavengers that donate electrons from their own molecules to neutralize the radicals. However, free radicals can multiply beyond the body's ability to complete or heal them naturally. For example, exposure to radiation, including that from the sun and from security x-rays, can accelerate the production of free radicals beyond the body's ability to keep them in check. More commonly, a diet high in fats creates free radicals.

Unchecked, the formation of free radicals generates a chain reaction. As each molecule impacts on more and more cells, more and more free radicals are produced, causing more and more unnatural biochemical processes.

Free radicals attack the membranes of cells in the body, causing the elements of the cell to attack each other, leading to imbalances in the levels of calcium and other problems. Ultimately, free radicals interfere with the DNA, or genetic coding, of the cells, which can lead to mutations and resulting problems, including cancer or birth defects in future generations. Free radicals overtax the immune system and keep it so busy fighting the breakdown of the body's own cells that it cannot fight off viruses, microbes, and infections that attack it from without.

As we saw in Chapter 3, the flavone glycosides in *Ginkgo biloba* neutralize free radicals. By keeping the cells of the heart and blood vessels strong, it also aids the lymphatic system, which is an important component of the immune system.

BLOOD PRESSURE

High blood pressure, or hypertension, is a leading cause of heart disease, yet it has no obvious symptoms, and many people who have high blood pressure are completely unaware of it. (*See* High Blood Pressure: A Case History, on page 97.) Normally, pressure in the arteries rises and falls with activity. It is low during sleep and higher during vigorous activity. The reading from a blood pressure test gives two numbers. The first number is called the *systolic blood pressure,* which is the greatest amount of pressure exerted against the walls of the arteries at the moment the heart beats. The second number, called the *diastolic blood pressure,* reflects the lower amount of pressure, when the heart valve is closed and the elastic arteries squeeze the blood onward through the body. A normal blood pressure reading is usually defined as is 120/80 ("120 over 80") or less—in other words, a systolic pressure of 120 millimeters of mercury or less, and a diastolic pressure of 80 millimeters of mercury or less. Actually, acceptable blood pressure levels vary with age. Healthy young adults have a pressure of about 110/75. Healthy people around age sixty have a pressure of 140/90.

Borderline high blood pressure falls in the range of 150/90 to 160/95. Hypertension is regarded by some physicians as blood pressure greater than 160/95 in repeated measurements. High blood pressure can damage the arteries by wearing them out. Approximately 50 million Americans suffer from this condition. There is often no recognized cause in any particular individual, although diet and heredity play a part.

By maximizing the effectiveness of adrenal hormones and the functioning of the adrenal glands (see page 94), ginkgo helps keep blood pressure constant by causing certain blood vessels to constrict when blood pressure falls below normal. It also relaxes the arteries and veins in times of spasm, and it stimulates them in times of paralysis. On a daily basis, it promotes healthy blood pressure. *Ginkgo biloba* keeps the body not just alive, but vital.

HIGH BLOOD PRESSURE: A CASE HISTORY

A 40-year-old stockbroker named Stephen, who spent long, pressured hours at work, finally gave in to his wife's insistence that he have a check-up. He felt that he was perfectly healthy because he was able to meet the demands of his life and job, and could get by on very little sleep. In his opinion, if he needed to make any changes for the sake of his health, all he needed to do was lose just a little weight. Happily for him, he had never smoked in his life.

Stephen went to his wife's general practitioner without a qualm, although he did wince when the nurse drew blood. When the exam was over, he dressed and went to the doctor's wood-paneled office to discuss the results.

To Stephen's surprise, he learned that his blood pressure was shockingly high, so high that his doctor would not allow him to return to work. "You could have a stroke at any time," the physician said. "I am giving you a prescription, and I want you to take it for three days. Then return here and I will tell you if it is safe for you to submit yourself to any strain." When he finally was allowed to return to work, Stephen was told that he could not spend more than eight hours at the office and that he had to stick to a regular, reduced schedule, which included eight hours of sleep nightly.

He and his doctor eventually managed to bring his blood pressure under control, although they could not pinpoint a reason why he had developed such severe hypertension. Based on promising studies, I think *Ginkgo biloba* could have helped him avoid his problem in the first place.

LEG CRAMPS

Recurrent leg cramps can put a severe limit on an otherwise active life. (*See* Claudication: A Case History, on page 99.) Narrowing or inflammation of the artery wall—a condition called arteritis—reduces blood flow. In the lower extremities of the body, it results in leg pain, especially when walking. People who fly a lot, including pilots, flight attendants, and frequent travelers, are susceptible to leg cramps because the recirculated air in pressurized airplane cabins may not supply sufficient oxygen for the muscle cells. People who stand on their feet a lot, such as cooks, dentists, and hairdressers, are subject to stress to their lower extremities, which may lead to leg cramps.

In a study done in the mid-1980s, thirty-six smokers who were experiencing leg cramps were given *Ginkgo biloba* for sixty-five weeks. For the first twenty-four weeks, researchers compared them with thirty-five subjects who took a placebo. Then the subjects were given the choice of abandoning the treatment or continuing the treatment with follow-up at regular twelve-week intervals. Subjects who continued treatment reported that after twenty-four weeks, or six months, *Ginkgo biloba* therapy gave significant pain relief and allowed them to walk farther and more comfortably. This improvement continued throughout the sixty-five-week period of the study.

Studies show that ginkgo is effective in approximately 75 percent of cases studied in which patients complained of debilitating cold, numbness, or cramping in their legs. A study published in *La Presse Médicale,* September 26, 1986, found that three to six months after treatment with 120 milligrams of *Ginkgo biloba* extract daily, people were able to increase their pain-free walking distance because the flow of blood to the limbs had been increased.

Apparently, ginkgo eases leg cramps not only because it increases circulation, but also because it improves the metabolism of cells in damaged tissues. Tissues that have been starved of nutrients for a number of months begin to generate free radicals. Ginkgo cleans up free radicals. The result is that the blood can bring the right amount of oxygen to feed

CLAUDICATION: A CASE HISTORY

A popular writer named Betty made her reputation with her stories of people she met in New York City over the course of several decades. She had no greater pleasure than to end her work day by taking long walks through her city. She sometimes roamed as much as sixty blocks, and she took great delight in going to Central Park near her apartment, which is a popular gathering spot for people and their dogs.

When she was about sixty-five, her legs began to tire more easily than they had in her younger years. Betty did not find this unusual. She knew that growing older caused a few physical problems, but she refused to give in to it. She thought that she could train her mind to overcome the ache, and then the pain. She would walk a few blocks and rest against a car or sit on a park bench when the pain in her calves became too much. After a while she could usually resume walking without discomfort, but over the course of a few months her courage and determination were not enough. The pain was too much for her. Slowly, she shortened her walks, until all she was able to do was go downstairs, sit in front of her building, and watch other people walk past.

Her problem was leg cramps, what doctors call claudication. How did leg cramps get such an unlikely name? Tradition has it that the syndrome was named for the Emperor Claudius, who was famous in ancient Rome for his limp. Doctors use the term *claudication* to refer to cramps in one or both legs that develop when one walks and that may eventually become a limp. It is a circulatory problem, and its cause is usually a blocking or narrowing of the walls of the arteries.

the cells. This has been shown by measuring the consumption of glucose and oxygen in the cells.

In Germany, ginkgo is approved as a treatment for leg cramps and numbness due to poor circulation. The distinguished British medical journal *Lancet* reviewed data on *Ginkgo biloba* extract in 1992 and noted that early research on the herbal extract as a treatment for leg cramps due to poor circulation was "promising."

SUMMARY

In this chapter, we have considered the workings and function of the entire circulatory system, and how *Ginkgo biloba* improves and maximizes the health of the heart, the blood, and the blood vessels. In the blood in particular, we see the benefits of the herb's ability to clean up or scavenge free radicals.

The health of the blood vessels is crucial to top performance of the human body. We would not put fine wine or pure water in a rusty container and drink from it. It makes no more sense to allow our blood vessels to deteriorate and carry our vital blood. *Ginkgo biloba* has been shown to benefit the blood vessels.

Promoting the health of the blood vessels counteracts, not surprisingly, a constellation of medical conditions. Because *Ginkgo biloba* counteracts hardening of the arteries, it eases leg cramps that plague older people and those who must stand on their feet for long periods of time. It also can improve blood pressure and overall cardiovascular health.

In the next chapter, we will go into specifics of how ginkgo's ability to improve circulation affects specific organs of the body, such as the eyes and the ears. French and German studies show that *Ginkgo biloba*'s ability to improve blood circulation can help offset a certain type of deafness. Poor circulation to the retina over the course of time dims vision, which is a major reason why elderly people experience vision loss. Scientists are finding that damage to the retina, particularly that caused by damage from years of ordinary exposure to sunlight, may be arrested by this age-old herb.

CHAPTER 6

Ginkgo and the Senses

We rely on our senses for everything we know about the world. In this chapter, we will look at *Ginkgo biloba*'s ability to enhance and improve the effectiveness of three of our essential senses: the sense of sight, the sense of hearing, and the sense of orientation, which involves equilibrium and location, both of which are governed by the inner ear.

Because *Ginkgo biloba* fights free-radical damage, it can prevent many eye disorders and treat these conditions in their early stages. *Ginkgo biloba* extract may delay the onset of these conditions and allow people to continue to enjoy normal, correctable vision for a much longer time than they would be able to if they did not take the extract. The herb's ability to regulate fluids in the body helps improve hearing, particularly in the elderly. Because the inner ear regulates balance, improvements in the circulation of fluid in the ear affects equilibrium as well.

To learn about *Ginkgo biloba*'s collection of benefits to the senses, let us start at the head, with the encouraging impact *Ginkgo biloba* extract has on the eyes and ears.

GINKGO AND VISION

Vision involves the detection and interpretation of the color, form, distance, depth, and dimensions of objects. It begins when light waves hit the retina of the eye. The lengths of those waves, and their amplitude, are the raw data of vision. Of the many factors that influence vision, the condition of the eye itself is as important as any. Of course, our ability to see also depends to a great extent on our heredity. If our forebears wore eyeglasses, so—most likely—will we, beginning sometimes in childhood. But as any grandmother will tell you, we have to take care of our eyes. Often, we appreciate this only when it is too late. When we are young, middle-aged people at restaurants who look first through the tops of their eyeglasses, then through the bottoms, and finally remove their glasses and pull the restaurant menus to their noses to read them properly are figures of fun. We imagine that we will never let ourselves get into such an awkward position, but time teaches us otherwise. The real difficulty of seeing comes home to us only when neither the eyeball nor the corrective lens can make a proper adjustment, when our "seeing devices" no longer work the way they used to.

Visual acuity is a measure of the resolving power of the eye. Anyone who has had an eye examination knows that acuity is usually measured by having a subject read letters of varying sizes from an "eye chart" at a distance. In the United States, the measured acuity is expressed as a fraction of what a person with normal vision would be able to see clearly at twenty feet. Thus, normal vision is 20/20. If your visual acuity is 20/60, you need glasses. If your best vision is 20/200 or worse (with glasses), you are considered legally blind.

There are differences between vision for close things and for distance. The normal eye has no trouble detecting distant objects, but it has to adjust, or "accommodate," for nearer objects, like a printed page. A six-year-old can see objects or text clearly at distances as close as 2.5 inches in front of his or her face, but with time the lens of the eye hardens and the eye's ability to focus on near subjects diminishes. The average thirty-year-old person can see clearly at a distance no nearer than six inches, and a typical sixty-year-old cannot clearly see something sixteen inches in front of him or her, making it difficult to understand such things as the menu in a restaurant, for example.

Obviously, if someone has been wearing glasses, ginkgo will not make his or her vision 20/20, but depending on the cause of the problem, it might improve vision to, say, 20/40. That is not a total victory, but such a gain could transform a person's life.

Because *Ginkgo biloba* fights free-radical damage, it may prevent eye disorders such as macular degeneration and damage triggered by diabetes, or it may treat these conditions in their early stages. *Ginkgo biloba* extract may delay the onset of these conditions and allow people to continue to enjoy normal, correctable vision for a much longer time than they would be able to if they did not take the extract.

The Structure and Operation of the Eye

The relationship between vision and the eye is something like the connection between the mind and brain. The eye is an organ, and vision is the faculty that depends on that organ, although the actual act of seeing—that is, making sense of the data that the eye gathers—is really done by the brain. The role of the eye is to send the vibrations of light into nerve impulses on to the brain. More than anything else, the eye is like a camera. It receives an outside image on its retina, much as a camera receives an image on film.

The eye is basically shaped like a sphere. The outer part of the eye consists of three layers, the most important of which is the light-sensitive retina. On its route to the interior of the eye, light travels through the cornea, a tough membrane; the aqueous humor, a clear, watery fluid; and the lens. A ring-shaped muscle around the eye focuses the lens by changing its shape. The iris, the colored part in front of the lens, opens or closes to regulate the amount of light admitted to the eye. The retina is a complex structure of light-sensitive nerve cells. It receives the images that come through the lens and the cornea. The nerve cells convert light into energy impulses, and other cells send these impulses to the brain via the optic nerve. The light-converting cells are of two types: rods, which are most efficient in dim light, and cones, which are most efficient in bright light. Directly behind the pupil is the area of the retina that is called the macula, or *macula lutea*. The macula is the area of the eye that is responsible for our most acute vision, including the ability to read small print and distinguish other fine details.

Declining Vision

At any age, a person's vision can weaken, either because the eyeball flattens, losing its round symmetry, or because the eye muscles stiffen and make focusing difficult. The most common results of these changes are *myopia* (nearsightedness) and *hyperopia* (farsightedness).

Over the course of time, poor circulation to the retina also dims vision, and this is a major reason for the loss of vision that most elderly people experience. Improved, carefully calibrated instruments have enabled us to detect and measure vision loss and other forms of degeneration in their earliest stages.

The most important effect of *Ginkgo biloba* on the eye is its effect on the retina. With all the "nervous" activity occurring there, the retina requires a steady supply of oxygen and glucose. The retina is particularly vulnerable to damage from free radicals. The tissues of the retina are rich in polyunsaturated fatty acids, which help in the transmission and conversion of energy that these hard-working eye cells perform. However, as we have seen, fatty acids are particularly attractive to damaging free radicals.

Ginkgo biloba is a scavenger of free radicals, and it improves circulation of the blood and delivery of glucose to all the tissues of the body. These capabilities of ginkgo account for its beneficial effects on the retina. Studies in Europe have confirmed these effects.

In the early 1990s, German scientists used the Octopus 2000 P, a system designed to study the health of the retina, to examine twenty-five people with a median age of seventy-five. They found that those taking a dose of 160 milligrams of *Ginkgo biloba* per day showed improved vision four weeks after treatment began. They reported that subjects showed "a significant increase in retinal sensitivity." A second group took 80 milligrams of ginkgo daily, half the dosage taken by the first group. Their vision improved after the dosage was increased to 160 milligrams. The more damaged the tissues, the greater the effect, while the healthy areas of the retina showed little change—another example of *Ginkgo biloba*'s having few or no side effects. Scientists determined that their test showed not only that *Ginkgo biloba* was effective in improving vision, but that damage to vision caused by poor circulation could, in large measure, be reversed. They reported that "damages to the visual field by chronic lack of blood flow are significantly reversible."

Researchers at the Institut Henri Beaufour in France fed rats *Ginkgo biloba* for ten days, then deprived their retinas of oxygen. They reported that *Ginkgo biloba,* used alone or in combination with drugs, enabled the tissues of retinas to withstand damage from oxygen deprivation.

A study done at the University of Leipzig also reported that *Ginkgo biloba* helps to protect the eyes. Scientists tested levels of enzymes that are released when the retina experiences inflammation or injury. Eighteen rabbits were fed 40 milligrams of *Ginkgo biloba* (EBb 761) intravenously each day for three weeks. When the researchers compared their retinas with the retinas of a dozen rabbits that had not received the extract, they discovered markedly reduced levels of the enzymes that would have indicated a problem in the rabbits that had taken *Ginkgo biloba*. Their 1994 report in *Ophthalmic Research* concluded, "*Ginkgo biloba* extracts may have a significant therapeutic value in cases of retinal damage."

Colorblindness

There are various forms of colorblindness. In some cases, its symptoms are that certain colors cannot be discerned. In others, there is an inability to distinguish between different shades of the same color. In still other cases, which are rare, the entire world is seen in shades of gray.

Colorblindness is usually the result of a fault in the retina or optic nerve, an inherited trait that is passed from mother to son. Females usually have genes that counteract the tendency. Certain degenerative processes affecting the retina can cause people to have trouble distinguishing between yellow and blue. This condition is sometimes called the "Swedish flag" syndrome because that flag has a yellow cross on a blue field. People with the syndrome cannot see the pattern. Another form of colorblindness, called toxic amblyopia, is caused by poisoning related to drugs, alcohol, or tobacco.

Aging brings on another type of color-vision deficiency. The ability to see and appreciate colors improves and peaks in a person's thirties and, like so many other types of sensory acuity, begins to decline after that. For some, this decline is merely an inconvenience that may cause minor sadness or annoyance. For others, it is a tragedy. The great painter Georgia O'Keefe, known for the beautiful desert paintings inspired by her home in New Mexico and for her vivid paintings of flowers, particularly orchids, lost her ability to see colors in her old age. Ever creative, she turned to pottery as a solution.

Here again *Ginkgo biloba* holds promise. A study reported in 1988 in the *Journal Français d'Ophthalmologie* revealed that the ability to distinguish colors improved in twenty-nine subjects who took *Ginkgo biloba* for six months.

The Effects of Diabetes on The Eye

Diabetes is a still-mysterious disease that is often inherited. It makes the body unable to store or use glucose, the body's chief source of energy, and it affects the blood vessels, though we do not yet understand how. It results in a high risk of atherosclerosis, hardening of the arteries, high blood pressure, and other cardiovascular disorders.

Blood vessels of the retina are particularly at risk for those with diabetes. Focusing on this problem, French researchers studied early damage to color vision caused by diabetes. The subjects selected for the study were unable to distinguish blue from yellow. The ability to distinguish colors in subjects who took *Ginkgo biloba* extract improved, whereas the condition of those taking a placebo worsened, as reported in the 1988 study cited above.

Diabetes can also damage the retina of the eye by causing the basal membrane of capillaries in the retinas to thicken. Scientists at Clermont-Ferrand reported in *La Presse Médicale* of September 26, 1986, that diabetic rats treated with *Ginkgo biloba* showed improvement to their retinas over the course of treatment. *Ginkgo biloba* was then administered to rats before the onset of diabetes, and they demonstrated an ability to resist retinal damage caused by diabetes.

Macular Degeneration

The macula, the most sensitive part of the retina, is vulnerable to a progressive, painless, but heartbreaking disorder called macular degeneration, which strikes many people in their old age. In the United States, it is the leading cause of legal blindness among the elderly.

Over time, a substance called lipofuscin can accumulate in the retina. Also known as "old-age pigment," lipofuscin is a lipid-containing golden-brown pigment that is a residue from the degeneration of cells. An accumulation of lipofuscin in the retina is associated with macular degeneration.

In the early stages of macular degeneration, the layer of insulation between the retina and the blood vessels behind them begins to break down. Fluid begins to leak from the blood vessels. Then scar tissue forms in the macula, leading to a dot in the center of the macula—and a corresponding blind spot in the field of vision—that grows progressively larger. Sometimes, treatment with a laser can seal off the leakage, but in most cases the disease is irreversible.

Because the macula is in the center of the retina, macular degeneration causes vision loss in the center of the visual field. This makes reading at first difficult, then impossible.

This is a tragedy for intelligent people who like to follow what is happening in the world around them through newspapers, or who love to lose themselves in the world of books. While they retain their "side" or peripheral vision, people with macular degeneration are still able to move about, but as the condition progresses they become more and more confined. Obviously, driving becomes impossible. They lose their ability to get around, to take care of their errands, do their own grocery shopping, and otherwise lead an independent existence. Many people with macular degeneration find themselves depressed by their loss of freedom.

Researchers have concluded that ginkgo may prevent macular degeneration, or at least slow its advance. In the mid-1980s, French scientists studied ten subjects with macular degeneration, giving half of them *Ginkgo biloba* and the other half a placebo. Those taking the ginkgo extract showed an improvement in their ability to see objects at long distances. Despite the small sample of patients, these results are statistically meaningful. This study was reported in the September 26, 1986, issue of *La Presse Médicale.*

It is quite interesting to see that ginkgo extract is accepted in Europe as a treatment for age-related macular degeneration. I hope that people whose eyes have been exposed to strong sunlight will find out about ginkgo in time, because the earlier a person takes it, the more effective gingko will be in combatting the ill effects of sunlight. Once a person has lost more than 20 percent of his or her vision to macular degeneration, it is not going to come back.

GINKGO AND HEARING

While the inability to see clearly is a nuisance and sometimes a tragedy, for some reason many people experience the inability to hear properly something of a disgrace. The range of hearing varies from person to person, just as the range of vision does. Perhaps we accept vision problems more readily because they are so common and because even children wear glasses. Hearing, like sight, is prone to degenerate with age, so that one-fourth of those over sixty-five need hearing aids. However, because of the social stigma, many people would rather struggle to pretend to have perfect hearing than have their hearing tested and, possibly, improved.

The Structure and Operation of the Ear

The ear itself is an elegant structure. The outer and middle ear collect sound waves, and the inner ear processes them, converting them into nerve impulses that travel via the auditory nerve to the brain, where they are evaluated. The inner ear also helps us maintain balance.

The outer ear consists of the visible part (called the *auricle* or *pinna*) and the ear canal that leads from it. It leads to the eardrum, a membrane that separates the outer ear from the middle ear. The eardrum vibrates in response to sound or pressure changes. Behind the eardrum lies the middle ear, a small cavity containing tiny bones called the hammer, anvil, and stirrup (or, in Latin, *malleus*, *incus*, and *stapes*). In the middle ear, the vibrations of air that have been caused by a sound and received by the outer ear are transferred to the liquid medium of the inner ear. The inner ear lies deep within the skull and is a maze, or labyrinth, of passages. The front part is the *cochlea*, a passage wound up like a snail, which contains the sensory receptors for hearing. Behind the cochlea lie three semicircular canals that contain receptors for the senses of equilibrium and position.

Hearing Loss

The ear is a delicate sense organ; it is sensitive and subject to injury and degeneration. As we live longer, we will experience more hearing problems due to the normal aging process. Two of the problems time brings are viral infections and a buildup of fluids in the inner ear.

Causes of hearing loss can be as simple as earwax blocking the outer ear canal. Sometimes hearing loss is caused by a buildup of fluid in the ear. Some deafness is conductive; in other words, the problem lies with the transmission of sound from the outer to the inner ear. Other deafness is due to a defect in a structure of the ear itself, such as the inner ear or the acoustic nerve. The inner ear can be damaged by long or sudden exposure to loud noise. Certainly baby boomers have experienced special stress to their hearing because of the loud rock music they enjoyed, particularly in their younger years. Certain drugs, particularly antibiotics, and even aspirin and diuretics can damage the delicate functioning of the ear.

The more we know of the elegant design of the human body and its intricate, ingenious systems, the more we want to do our part to keep them in good order. We know we can assist our eyes by reading in good light and by protecting ourselves from sun damage by wearing sunglasses. Often, it seems there is little we can do for the ears—other than avoiding being too near to loudspeakers. But there is evidence that we can protect their complex cells and the fluid of the inner ear by taking *Ginkgo biloba* as a preventive measure.

A French study reported in *La Presse Médicale* on September 25, 1986, showed that *Ginkgo biloba* can help offset acute cochlear deafness, which results from damage to the cochlea, the structure that transforms the vibrations of sound into nerve impulses sent to the brain. Ginkgo is effective against this type of deafness because of its ability to improve blood circulation in the ear. The patients in the study were in the early stages of acute cochlear deafness. Half the subjects were given 320 milligrams of *Ginkgo biloba* daily, and the other half were given a type of drug called an alpha-blocker. All of the subjects experienced some improvement, but 52 percent of those taking the herbal extract showed substantial hearing improvement, versus 35 percent of those in the group that took the alpha-blocker.

Another study, also reported in the September 25, 1986, issue of *La Presse Médicale*, compared the effectiveness of two active drugs in treating patients who became deaf suddenly, exposed either to a burst of sound, as at a rock concert, or to a sudden drop in pressure, such as that caused by the failure of a pressurized cabin in an airplane. None had been exposed to bacteria or had other causes for their deafness. Ten patients in the trial were given 80 milligrams of *Ginkgo biloba* in drinkable drops twice a day. Ten others were given a standard alpha-blocker twice a day. No other drugs were given.

The subjects were hospitalized for ten days and were tested every other day. A month after they had been released, they were checked again. The alpha-blocker group had three very good results, three "fair" results, and four failures. The ginkgo group had seven excellent results, one "mediocre," and one failure. Both groups improved, but those taking ginkgo improved much more noticeably. Researchers linked ginkgo's superior results in treating the ear to the complexity of the *Ginkgo biloba* extract. They said that it improved circulation in the tiny blood vessels that had suffered from oxygen deprivation and it protected cells and blood vessels in some unspecified way. Hearing was most improved on the tenth day of treatment. Nonetheless, researchers recommended that people using ginkgo for hearing problems take *Ginkgo biloba* extract for thirty days to show hearing improvement.

Tinnitus

The only thing worse than not being able to hear sound is never to experience silence. Ringing in the ears, a condition known as tinnitus, is a particular problem for the older people. It involves not only ringing, but buzzing, hissing, and whistling sounds. Normally, the acoustic nerve in the ear transmits external sound waves. In tinnitus, it responds to some sort of stimulus or triggering in the head. It is considered a symptom of poor circulation in the brain, and it is a malady for which *Ginkgo biloba* has demonstrated clear effectiveness as a treatment.

In a landmark study of the effects of aging on the brain reported in *La Presse Médicale* on September 25, 1986, French researchers included tinnitus among the "cerebral disorders" that *Ginkgo biloba* treats effectively. Researchers who focused solely on ringing in the ears came up with even more encouraging results. In the mid-1980s, Europeans conducted a multi-center, three-month, double-blind study of 103 patients who had had tinnitus for less than one year. The patients in one group were given 320 milligrams of ginkgo daily for thirty days; those in the control group were given a placebo. The severity and intensity of the condition varied among the patients, but all who received *Ginkgo biloba* for a month improved much more than those taking the placebo. Their tinnitus was either minimized or improved.

GINKGO AND EQUILIBRIUM

The ability to remain upright is a complex process that involves the eyes, which tell us where we are relative to what is around us; the sensory nerves, which inform the different parts of the body where they are relative to each other; and the three canals in the inner ear. The brain, of course, monitors and processes this information and prompts the body to act on it. Much of the information that the brain employs in this process comes from the inner ear.

As much as an organ of hearing, the ear is an organ of balance. It enables us to move without falling and to stand erect. The three semicircular canals that are in the rear area of the inner ear regulate equilibrium. They are set at right angles to each other so that they can detect movement on any plane. They contain hair cells that are surrounded by a jelly-like fluid and connected to nerve fibers that are connected to the central nervous system. The fluid moves when the head moves, and as the fluid moves it disturbs the hairlike cells, which activate signals in the nerve fibers. Some of the hairs and nerve cells are affected by gravity and acceleration, others by the position and movement of the head up and down, forward and backward, and to the left or right.

Dizziness, a feeling of being unsteady or lightheaded, is usually mild and brief. In serious cases, however, it is a sign of vertigo, a more serious problem, a sensation that one's surroundings are spinning. Any disruption in the semicircular canals that regulate balance, such as a change in the circulation of fluid, can cause vertigo. People with atherosclerosis—a thickening and narrowing of the wall of an artery—can experience vertigo if they so much as move their heads. Here, *Ginkgo biloba* is an appropriate treatment because it keeps the arteries healthy and promotes blood flow so that the brain can properly receive and evaluate the information it receives, rather than creating the sensation of being in a whirling room.

The French have done most of the research on the question of *Ginkgo biloba* and equilibrium. In a three-month, multicenter study reported in the September 25, 1986, issue of *La Presse Médicale,* researchers gathered sixty-seven patients who had had a persistent problem with balance for less than three months. Thirty-four patients were given 160 milligrams of *Ginkgo biloba* each day, and the rest of the group received a placebo. Scientists evaluated the subjects on a visual scale and by standard tests, and then rated them for major improvement, slight improvement, stable condition (meaning no change), or deterioration of balance. After ninety days, 74.7 percent of those taking ginkgo had improved. This rate of improvement is particularly dramatic in comparison with the rate of improvement for those who did not take *Ginkgo biloba* extract. Only 18.3 percent of that group improved.

Balance was restored slowly in the ginkgo group. Improvements were noticeable after thirty days, and were significant after sixty days. The big improvement, however, was seen after ninety days. Vertigo disappeared completely in sixteen patients who had taken ginkgo, compared with only six patients who had taken the placebo.

A 1995 study done on cats at the University of Aix-Marseilles in France showed that animals given *Ginkgo biloba* recovered their sense of balance and ability to move properly more rapidly and more completely than those that were not given the herbal extract.

Another French study reported in *La Presse Médicale* on

September 25, 1986, showed that *Ginkgo biloba* extract helped to end or ease problems with balance. Scientists investigated its impact on seventy people who had recently, suddenly, and inexplicably developed problems with their balance. After a double-blind study at three different centers extending over a three-month period, researchers found that ginkgo extract had a statistically significant impact. Almost half of the patients—47 percent—were completely rid of their symptoms, versus 18 percent of those who did not take *Ginkgo biloba* extract.

Whenever people notice a problem with balance, they should take *Ginkgo biloba* extract. The herb should not only help to clear up problems with equilibrium, but should help to moderate the underlying problems causing it.

SUMMARY

In this chapter, we have taken a look at a cornucopia of benefits offered by *Ginkgo biloba* extract. We have reviewed promising research that shows that ginkgo keeps the cells of the retina of the eye healthy, and may prevent macular degeneration, a type of blindness that plagues elderly people. It may successfully treat the problem in its early stages. Anyone who fears vision loss should be alert to the emerging research on *Ginkgo biloba* and the health of the eye.

Ginkgo has been proven effective in treating tinnitus, or ringing in the ears, and problems with balance. Ginkgo's ability to improve the health and functioning of the blood vessels of the brain, and to keep them functioning properly, enables ginkgo to help with these conditions. Problems with balance or equilibrium are very frequent, and most of the time they cannot be attributed to a single cause. Sometimes doctors administer extensive, expensive tests and come up with no cause. But if you start treatment with *Ginkgo biloba* early and take it for three months, you may experience an improvement.

As has been demonstrated over and over, *Ginkgo biloba* extract keeps cell membranes strong and cleans up damage that occurs in, and from, the fluids of the body. Indeed, so many and so significant are its potential benefits that *Ginkgo biloba* extract deserves to be as much a part of our health regimen as vitamin supplements.

CHAPTER 7

Ginkgo and Sexuality

In this chapter, we take up maladies that affect sexuality and the reproductive system. Ginkgo is able to combat certain forms of premenstrual syndrome (PMS), particularly the swelling and aching that trouble many women every month. Also, just as it renews blood vessels of the brain and legs, so it can revitalize the blood vessels of the penis, helping men who suffer from impotence.

PREMENSTRUAL SYNDROME

For too long in the history of modern medicine, women's health problems were ignored or minimized. Because women were seen as emotional creatures, all of their physical problems were seen to have an emotional basis. Problems surrounding menstruation, in particular, were said to be "idiopathic." The dictionary will tell you that the term *idiopathic* means "peculiar to the individual." In fact, it means that the doctor doesn't know what is causing the problem. The medical establishment has until recently tended to view women as always whining and complaining about imaginary discomforts.

In the earlier chapters of this book, we have noted many

instances that illustrate the intricate interactions that take place to bring about bodily processes we take for granted. Menstruation, not surprisingly, is the result of a complex series of interactions. At the beginning of the monthly menstrual cycle, estrogen—the so-called female hormone—causes the endometrium (the lining of the uterus) to thicken and prepare for the possibility of pregnancy. Ovulation, the release of an egg, occurs two weeks later. It is accompanied by an increase in production of the hormone progesterone, which causes the cells of the lining of the uterus to swell with fluid so that, if fertilized, the egg will attach to it.

If the egg is fertilized by a sperm and implants in the uterine wall, pregnancy results. If not, then the production of estrogen and progesterone decreases. The spongy, blood-rich lining of the uterus sheds, usually over the course of several days, in the menstrual—or monthly—discharge.

Premenstrual syndrome begins a week or two before menstruation. It affects an estimated 90 percent of all women at some point in their lives, a few so severely that they are incapacitated and cannot go about their daily tasks. The most frequent symptoms affect mood; they include irritability, nervousness, depression, and fatigue. However, there are also significant physical symptoms. A premenstrual woman's body retains fluid. Her breasts are swollen and tender. She has headaches, backache, and cramps. Her legs may swell as well. Various treatments, from vitamins and minerals to hormones and diuretics, have been prescribed, and help some women reduce swelling. Ibuprofen, which was developed to help people with arthritis, helps alleviate abdominal pain.

Fortunately, one doctor, R. Mach of Geneva, Switzerland, took the swelling and water retention that accompany menstruation seriously, and in 1955 he uncovered their cause. Unfortunately, years passed before his findings were made available. For decades, the medical establishment thought his work was too mundane because it related only to women.

Dr. Mach was a kidney specialist who studied the movements of water in the body. He studied what happens if we either do not eliminate water from our tissues, or eliminate it so fast that we get dehydrated. He described the weight gain in premenstrual women due to water retention as rapid and occurring only during the day. He also found that it was labile, or unstable, changing from day to day. Water retention affects the skin, the tissues under the skin, and eventually the muscles of the legs and the feet; it can amount to a gain of up to three pounds a day.

Interestingly, in addition to premenstrual women, people who stand for long periods time—such as waitresses, flight attendants, dentists, and salespeople—also experience this problem. Fluid retention is worse when the temperature is warm or where floors are heated. It lessens when the afflicted person is lying in bed, because when one is lying down, one produces more urine.

In some extreme cases, so much water accumulates that the legs resemble those of elephants. Dr. Mach concentrated on a group of women who complained of thirst but did not urinate very much. They were tired and complained of shortness of breath. They were constipated and had headaches. The physical processes that lead to these symptoms are easy to understand. Retention of water and salt (or sodium) caused their capillaries to leak and to allow water to seep out into the tissues. Distension of the capillaries and higher pressure within them led to the leakage.

Mach's investigations of patients with extreme fluid retention revealed that they produced too much of a hormone made in the adrenal glands, called aldosterone, and that they were slightly deficient in the production of progesterone. Both of these hormones are produced mostly in the second half of the menstrual cycle.

Most women do not go to doctors about their menstrual problems, partly because they have been so badly treated in the past. They do tend to go to pharmacists and get diuretics and laxatives. Actually, these two types of drugs fight the hormonal system, so that it ends up producing *more* aldosterone, which was the cause of the problem in the first place. The use of laxatives also leads to a loss of potassium and other valuable minerals. And people who take laxatives regularly only become more constipated, because the effect of laxatives is temporary, and when one takes a laxative more frequently, the body no longer responds to it. Further, if you stop using salt because you don't want to retain water, your body will fight to retain salt because it requires sodium.

Obviously, this is a vicious circle. The first step in treating women with premenstrual syndrome is to convince them to give up laxatives and diuretics. Most resist this, because they think that if they avoid diuretics, they will end up looking like the Michelin man.

There is some evidence that *Ginkgo biloba* should be added to the arsenal of available medications to combat water retention, according to a report in the September 25, 1986, issue of *La Presse Médicale*. Researchers at the Henri Mondor hospital in Créteil, France, gave a battery of tests to a group of women who had suffered from severe problems with water retention for eleven years. Particularly important among these tests was the Landis test, in which patients are injected with albumin (a type of protein) labeled with radioisotopes that can be tracked through the body. The subjects in the study were divided into two groups. In the first group of women, the researchers discovered that there was leakage of albumin from the blood vessels into all tissues, particularly in the legs. They treated these women for one to two months with *Ginkgo biloba* extract at dosages between 160 milligrams and 200 milligrams daily. Every one of these women achieved a normalization of the Landis tests—no more leakage from the blood vessels into the tissues. Not only that, but of the women in that group, three never experienced the problem again. Six had a major improvement and one had some improvement. One of them was no longer constipated. One dropped out of the study because she complained of palpitations and insomnia after a few days.

In a continuation of this study, five women who experienced even greater and more debilitating fluid retention were admitted to the same French hospital and given an intravenous infusion of 200 to 300 milligrams of *Ginkgo biloba* extract once a day for four to five days. Each lost between four and ten pounds that were due to water retention. Then the women were given *Ginkgo biloba* extract orally, and their problems were eased. We should add that these women, who had had such extreme problems, now wear special control stockings as a permanent treatment for their leg problems. Other than this, their treatment for water retention consists of a good diet with normal amounts of salt, sufficient rest, and *Ginkgo biloba*.

Since the results were so encouraging in women who suffered from extreme premenstrual syndrome, it is not surprising that *Ginkgo biloba* is helpful to those whose discomfort is more average. In the mid-1990s, French researchers studied women who reported that they had suffered congestive premenstrual syndrome (swelling or discomfort due to fluid retention) during one week each month for at least three months. The study involved 165 sexually active women between the ages of nineteen and forty-five, who were divided into an experimental group and a control group. For two months, from the sixteenth day of one menstrual cycle until the fifth day of the next one, the patients in the experimental group were given *Ginkgo biloba* extract, and the patients in the control group were given a placebo. The patients rated their conditions themselves, using a daily rating scale, and doctors evaluated them before and after their periods. The results indicated that *Ginkgo biloba* was effective against congestive symptoms of PMS, particularly breast discomfort and irritability. There were no side effects and there was no impact on hormonal balance.

IMPOTENCE

Impotence is the inability to have or to maintain an erection. Most men have this experience occasionally, but when it develops into a chronic pattern, it is a problem. Some 30 million American men suffer from it, but only one-tenth of them seek medical help. Many assume that the problem is hopeless; others are too embarrassed to expose their "weakness," or even to admit to themselves that they have a disorder. Even the Latin root of the word suggests the stigma attached to the condition; it comes from *impotentia,* which means "a lack of strength."

Many men accept impotence as yet another inevitable sign that they are getting older. It is true that men over sixty have a greater incidence of impotence than younger men. Almost 20 percent of men are impotent at sixty, and more than 70 percent are impotent at the age of eighty.

The American Medical Association's *Encyclopedia of Medicine* says that 90 percent of the time, when a patient complains of impotence, he has a transient psychological problem. Emotional stress, fatigue, and anxiety are usually responsible for short-term impotence. Depression and other negative psychological states—including bad feelings towards one's sexual partner—can also contribute to impotence, because a man's state of mind is the key to whether or not he can become aroused. However, long-term problems with impotence usually have a physical basis.

The word *impotence* is hardly vague, yet it is one word that covers a wide range of degrees. A man can be "impotent" for one night, for a week, a month, or even six months. Sometimes years pass, and if conditions in his life change, he can "function normally" again. In such cases, the problem is caused by mood, state of mind, or a response to external factors, such as stress, exhaustion, or low self-esteem.

Today, experts in human sexuality, called sexologists, say that most cases of chronic impotence in men of any age are physical in origin—caused by problems with blood vessels or the nervous system. In 90 percent of these cases, the problem is organic—a problem with the circulatory system or nerve damage.

Physicians—particularly urologists, who specialize in disorders of the penis, prostate gland, and testicles—can administer physical tests to determine whether long-term impotence is caused by a physical problem. The many possible physical causes of impotence include high blood pressure, diminished testosterone levels, side effects of medication taken for other conditions, and damaged blood vessels. Men who have intensely physical jobs are the most likely to experience injury and so, despite their macho image, police officers and firefighters are the most subject to impotence due to trauma. The most frequent physical cause of impotence is blockage to the flow of blood to the penis.

Once it is established that there is a physical basis for impotence, the question becomes, what is the best treatment? Surgical implants are one option. Another is a "vacuum constriction device," a pump composed of a cylinder the size of a test tube that is placed over the penis. The pump removes the air and draws blood into the penis to create an erection. An elastic band is then placed at the base of the organ to maintain the erection. Injectable drugs are yet another option. One, called Caverject, relaxes the muscles of the penis and improves blood flow. A newer drug, called Muse, is administered through a plunger.

Ginkgo biloba improves circulation to all the blood vessels, including those of the penis. There have been encouraging experiments indicating that ginkgo can help treat or prevent impotence due to restriction in the blood vessels of the penis. It is also showing promise as a treatment for sexual problems related to the use of commonly prescribed anti-depressant drugs, such as fluoxetine (Prozac) and sertraline (Zoloft).

Ginkgo's beneficial effects make sense when we consider the physical mechanism of an erection. When a man is sexually aroused, the nerves of the penis increase blood flow to that organ by 700 percent. This sudden rush of blood expands, strengthens, and hardens the penis into an erection. After the semen is ejaculated, or when the sexual stimulation passes, the blood drains away, and the penis returns to its normal, non-erect state. Because an erection requires not only arousal, but the action of the nervous and circulatory systems, anything that damages the nerves or restricts the blood vessels will hinder a man's ability to achieve an erection.

In a study reported in the *Journal of Urology* in 1989, sixty men who complained of problems in achieving and maintaining erections because of poor circulation in the arteries of the penis, and who had not been helped by injectable medication, were given 60 milligrams of *Ginkgo biloba* extract daily for twelve to eighteen months. By the end of the study, half of the subjects had regained the ability to have erections. They had been cured. Another quarter of them had improved blood flow in their arteries during the time they were studied.

In another trial, reported in the *Journal of Sexual Education Therapy* in 1991, fifty patients were given 240 milligrams of *Ginkgo biloba* extract for nine months. The group was divided into those who had previously benefitted from drug treatment for the problem and those who had not. All of those who had previously reacted positively to drug treatments showed improved blood flow to their penises in the first three months, and all of those had "sufficient" erections after six months of treatment. Those who had not previously benefitted from drugs experienced at least improved arterial flow at six and nine months.

Certainly, more research is needed. The idea that ginkgo treats impotence is based on tradition in Asian medicine. I am quite sure that it does help, because it increases the blood flow in the erectile zones of the penis, but whether it helps patients' attitudes or their bodies has not been proven. We know that ginkgo has no adverse side effects, so that if studies prove that it is even at least as effective as drug treatments for impotence, ginkgo will have an advantage.

As the public becomes more sophisticated and realizes that "sex problems" are not a sign of weakness or inadequacy, these problems will be discussed, brought to health care professionals, and treated. Relatively little research has been done on using *Ginkgo biloba* as a treatment for impotence, but all the existing evidence that ginkgo aids the blood vessels and circulatory system argue strongly for its potential in treating impotence, which causes needless embarrassment and unhappiness to so many men and those who love them.

SUMMARY

Ginkgo biloba, as we have seen, helps cell membranes stay

healthy and helps them resist leakage of fluid. It also promotes the health of the blood vessels, so it is not surprising that ginkgo has been effective for women experiencing menstrual difficulties and for men experiencing impotence due to poor circulation. It is too much of a stretch to say that this herb is guaranteed to help every man who experiences impotence or every woman who experiences monthly discomfort. However, I would recommend that men who have problems with impotence and women who have problems with premenstrual syndrome, including cramps, take *Ginkgo biloba* extract over the course of a few months and make note of whether or not their symptoms ease. Quite possibly, *Ginkgo biloba* will be an effective treatment.

CHAPTER 8

Ginkgo's Other Benefits

Ginkgo has so many beneficial effects that it seems somewhat unbelievable. These range from improvement of memory and circulation to relief of leg cramps and ringing in the ears. In fact, this grab-bag of benefits results from ginkgo's three basic actions: It protects the health of the cells and keeps them from breaking down under attack; it cleans up free radicals that set off damaging chain reactions; and it keeps the blood vessels strong.

In this chapter, we take up a range of other maladies that are improved by *Ginkgo biloba*. We have noted that because *Ginkgo biloba* scavenges free radicals, it fights damage caused by the radiation of the sun. In rarer but more frightening situations, *Ginkgo biloba* combats damage to the cells caused by disasters such as atomic bombs and nuclear plant malfunctions. It helps the white blood cells fight severe infection throughout the system, such as in septic shock. It also fights inflammation due to allergies by stopping the action that turns the immune system against itself. Because *Ginkgo biloba* fights inflammation, it can help those who suffer from two very different conditions: asthma and hepatitis. Asthma, which makes breathing difficult for many people, is inflammation of the airways. The inflammation that is hepatitis irritates and damages the liver.

We have seen that ginkgo counteracts platelet-activating factor (PAF). In addition to its role in stroke and heart attacks, platelet-activating factor triggers allergic reactions and asthma. It promotes inflammation and disorders of the central nervous system. These include injury to nerve cells, shock, inflammatory conditions, allergies, and bronchial constriction. PAF may contribute to asthma, the rejection of transplanted organs, shock, irregular heartbeat, kidney disease, and chronic problems due to head injuries.

GINKGO AND RADIATION

Since the beginning of time, there has been radiation on earth. Ultraviolet rays from the sun, cosmic rays from space, and the natural decay of some of the earth's minerals—such as radium—have meant that people have always been exposed to some radiation. Radiation is basically the transfer of energy from an atom through space. Sometimes radiation occurs in waves, like sound, and at other times as particles, such as the cosmic rays that come from outer space. Radioactivity is the emission of radiation that occurs when unstable substances, such as uranium, disintegrate and throw off alpha or beta particles or gamma rays.

In the last century, the amount of radiation that we are subjected to has increased. Dramatic events, such as the testing (and dropping) of atomic bombs, and accidents at nuclear power plants, account for a small, though significant, part of the increase. Carelessness in the mining and refining of radioactive materials, such as uranium and plutonium, has allowed radioactive materials to seep into our ground water. As a result, not only do we absorb radiation by inhaling radioactive substances from the air, we also absorb radiation when we eat or drink because radioactive substances are in our food and water. For example, in 1958, when sixty-four atomic tests were conducted in the United States, high amounts of strontium-90 were found in the milk of cows that grazed downwind of the testing site. A paper published in the *Journal of the American Medical Association* in 1984 revealed that Mormons who lived in southwestern Utah, downwind of a nuclear-testing site in Nevada, had high rates of leukemia and other cancers, despite the fact that, as a group, Mormons are known to live healthful lives.

We usually learn about the dangerous effects of radiation only when it is too late. For example, from about 1913 until the early 1930s, luminous watch dials and instrument panels were painted with radium. The women who painted these instruments were told to lick their paint brushes between strokes to keep the points sharp so that they could draw finer lines. When a significant number of these workers began to die of cancer, manufacturers reduced—but did not eliminate—the amount of radioactive materials they used for luminous dials.

The effects of radiation build up over time. Because the radioactive substance polonium-210 is included in fertilizers that are used on tobacco plants, cigarette smokers receive the equivalent of some 300 chest x-rays per year. Radioactive material never really disintegrates; it has what scientists call a half-life. In other words, it goes through a decaying process in which half of the original amount of radioactivity will remain after a certain period of time. For example, strontium-90 has a half-life of twenty-eight years. This means that a sample of strontium-90 will lose half of its radioactivity in twenty-eight years, but it does *not* mean that it will lose the other half twenty-eight years after that. In the second twenty-eight years, half of the *remaining half* of the radioactivity will be lost, so that at the end of fifty-six years, a quarter of the original radioactivity will remain. It will take 560 years before the radiation of a quantity of strontium-90 is too low to be detected.

Unfortunately, because of accidents at nuclear power plants and because of nuclear testing in the 1950s and 1960s, all of us may be at some risk of exposure to radiation. Because there are so many variables about radiation doses and effects, scientists have not estimated what the actual risk may be to any given individual. However, legal experts have been willing to make judgments based on data available to them. In 1984, a federal judge in Utah ruled that there was a preponderance of evidence that ten people had developed cancer because they had unwittingly put themselves at risk of exposure to radiation, because they were not advised of the risk presented by nuclear fallout in their state. Since the signing of the limited test ban treaty in 1963, fallout levels have dropped around the world, but probably not enough for safety.

Ginkgo biloba's astonishing ability to scavenge free radicals gives us a means to conteract the radiation we are exposed to. Dramatic proof of this comes from the aftermath of the Chernobyl disaster. I touched on this briefly in Chapter 3. Now let us look at this story in more detail.

On April 26, 1986, a poorly supervised experiment was being conducted in one of the four reactors at the nuclear power plant in Chernobyl, Ukraine, while the cooling system was turned off. These conditions led to an uncontrolled reaction and a steam explosion that blew the protective covering off the reactor. As a result, radiation spread across northern Europe as far as Great Britain. More than 100,000 citizens of what was then the Soviet Union were evacuated from the area around the site. Official reports stated that thirty-one people died as a result of the accident, but U.S. scientists estimated that the number of deaths that radiation would eventually cause would be much higher.

The entire population of the area was at risk, but most at risk were the workers who were called from various areas of the Soviet Union to shut down the damaged reactor. After their work at the reactor site was completed, they were given thorough medical examinations, including various blood tests, to assess the state of their health and the degree of injury they had suffered due to radiation exposure. Clastogenic factors were found in the blood of thirty-three of forty-seven people who worked to clean up and shut down the Chernobyl nuclear plant after it malfunctioned. Clastogenic factors are free radicals that kill cells by damaging their chromosomes. These factors are a sign of radiation exposure.

Thus, the workers were already showing signs of chromosome damage, putting them at high risk of developing cancer. Medical researchers went to work, looking for a way to halt the chromosome damage. There are some drugs that can be used in such cases, but these are dangerous if taken over the long term. Researchers experimented to see whether using *Ginkgo biloba*, a known antioxidant, could reduce the clastogenic factors. The appeal of *Ginkgo biloba* was that it had never demonstrated serious side effects and could safely be taken over the course of a lifetime.

In a French-sponsored study published in 1995, researchers gave 40 milligrams of *Ginkgo biloba* to the Chernobyl workers three times a day over a period of eight weeks. At the end of the treatment period, the level of clastogenic factors in their blood had been reduced to "control level," the level in people not exposed to exceptional amounts of radiation, and this level was maintained for at least seven months without further treatment. One-third of the workers who stopped taking *Ginkgo biloba* showed evidence of dangerous levels of free radicals twelve months after they stopped treatment, indicating that their systems began to produce free radicals again after *Ginkgo biloba* was withdrawn. However, even this group had been helped by *Ginkgo biloba*. Thus, *Ginkgo biloba* was effective in combatting the effects of radiation. Some individuals benefitted for a longer period than others, but the extract helped to reverse damage caused by free radicals for varying amounts of time in all cases.

It is to be hoped that no one will ever again be exposed to the levels of radiation experienced by the workers at Chernobyl. However, the sad fact is that the normal course of modern life exposes us to levels of radiation that are probably dangerous. Today, each individual is exposed to potentially harmful radiation, some of it naturally occurring and some of it the product of human advances that make use of radiation, including:

- Medical treatments such as x-rays, radiation therapy, and magnetic resonance imaging (MRI).

- Air travel (there is less protection from radiation in the atmosphere at altitude).
- Video-display terminals (televisions and computers).
- Building materials that absorb radioactivity from the earth.

On our jobs, in our homes, and certainly at the doctor's office, we are exposed to radioactivity. Even if the dose we receive each time is at a safe level, the total dosage builds up over time. *Ginkgo biloba* can help protect against the negative effects from this everyday exposure.

GINKGO AND SUN DAMAGE

Sunlight is known to cause skin cancer and may promote other cancers as well. German scientists studied how *Ginkgo biloba* affects the body's ability to cope with the sun's ultraviolet light, and whether ginkgo can block sunlight from altering cells in the body. They reported in 1992 that *Ginkgo biloba,* used in combination with Vitamins A and E and selenium, inhibited cell damage. French scientists studied the effects of ultraviolet light on lab animals. They reported in *Free Radical Biology & Medicine* in 1992 that if organ tissues were treated with *Ginkgo biloba,* they resisted attack by free radicals after being exposed to radiation from ultraviolet light.

Aging

The sun accelerates the aging process of the skin. Young skin is thick, with many layers of collagen, the curvy protein fibers that make the skin supple and elastic. Older skin, in contrast, is thinner and less flexible. It is more vulnerable to injury and it heals more slowly because its capillaries are weaker and because its collagen layers have thinned.

The sun causes worse damage to the skin than anything natural aging can do. It makes oily skin oilier, makes normal skin dry, and makes dry skin crack. The sun's rays are essential to life on our planet, but our skin thrives best with limited exposure to sunlight. Knowing that people liked the look of a "healthy" tan, doctors once tried to find how a person could have an attractive tan without endangering the skin. They recommended that people tan slowly, so that the skin could build up a layer of pigment that would protect against burning. However, once effective instruments were developed to measure the effect of sunlight on the skin, doctors learned that a tan offers little protection against the harmful radiation in sunlight. With a minuscule sun protection factor of two to three, a tan is less effective than suntan lotion. After a certain amount of exposure to the sun, skin begins to age, even if the skin is that of a sixteen-year-old girl. Sunbathing makes no sense for anyone, because any exposure to the sun has a harmful effect on the skin.

Why does sunlight hasten the aging of skin? One type of ultraviolet ray in sunlight, ultraviolet B (UVB), is the most damaging and the most responsible for negative effects to the skin, such as aging, burning, and cancer. These negative effects occur because UVB rays disrupt some light-sensitive proteins. The resulting chemical reaction makes the blood vessels redden and swell. This swelling, or edema, initially "pumps up" the skin, making lines disappear, but it is actually a sign of injury. The cells try to protect themselves by reproducing rapidly to provide an extra layer of skin to act as an umbrella shielding the skin from the sun. Meanwhile, ultraviolet A (UVA) rays activate pigment-producing cells called melanocytes to produce more melanin, or dark pigment—the tan we desire—as an additional shield against the sunlight's attack. Over time, unfortunately, UVA rays aggravate the damage done by UVB rays. Sun poisoning is largely the work of UVA.

While the tan may seem to fade, the melanin remains. The cells that contain melanin start to die and slough off, resulting in a loss of cells. Most "natural" aging of the skin is really sun damage. Just contrast the smoothness of your complexion with the smoothness of, say, your midriff. When one is young, the body replaces damaged collagen, but the ability to replenish it declines with time. Damaged collagen remains in the skin, and the skin loses its elasticity. It sags, causing bags and wrinkles.

Sun also damages elastin, straight protein fibers that stretch and that also give the skin its elasticity. UVB rays make elastin reproduce itself improperly, so that it has no resilience. Skin becomes flabby, with large pores and rough texture. These "bad rays" and the heat of the sun cause cell damage, which impairs blood circulation in the skin, making it more yellow and less nourished.

Here again, *Ginkgo biloba* helps. Its demonstrated ability to "clean up" destructive free radicals and its ability to strengthen the capillaries help protect the skin and heal it. *Ginkgo biloba* extract helps the body combat damage that takes place in the cells when we are exposed to sunlight. This herb is not only a medicine, it is a beauty treatment. Ginkgo helps keep skin beautiful.

Skin Cancer

The simple fact is that skin cancer seldom appears on skin that has not been exposed to the sun. The most elementary study of data from around the United States shows that people in New Hampshire, where winters are long and gray and rainfall is relatively high year round, have less skin cancer than people in southern California, where a rain shower in any month but January is practically a news event.

The incidence of skin cancer has soared 900 percent since the flapper era of the 1920s, when women discarded their parasols and began "bathing" in the sun. Mercifully, skin cancer is rarely fatal, but it does cause mental distress, run up medical bills, and, often, cause scarring.

GINKGO AND ALLERGIES

Allergies are excessive and inappropriate reactions of the immune system. The body becomes oversensitive to what is supposed to be a harmless substance and decides to fight it off. Outward signs of allergies include skin rashes, sneezing, runny noses, even vomiting. Sometimes various organs are affected and begin to malfunction. In allergic reactions, people's skin can redden or blister because they eat strawberries, or they can begin to sneeze and wheeze because pollen is in the air, or their stomachs can rebel unpleasantly just because they drank some milk.

These are complicated conditions, and researchers have not found why any one person develops an allergy, although often allergies run in families. Sometimes the environment sensitizes a person; living with a cat or dog can make a child hypersensitive to the dander in animal's fur, or eating a diet too high in dairy products can cause the immune system to overreact.

When one is allergic to a substance, the immune system produces antibodies that coat cells in the lungs, stomach, and upper respiratory tract. When one ingests or breathes the allergen again, these cells release various chemicals, which are what set off the symptoms of the allergy. Among these chemicals is histamine, which makes the blood vessels swell. Fluids then begin to leak into the tissues from the blood vessels and muscles can spasm. If you have seen a child red and swollen with hives, you have seen the action of histamine on the blood vessels.

Happily, *Ginkgo biloba* can help stop the biochemical reaction in the body that results from an allergy. Ginkgo slows the leakage of fluid from the capillaries that can be induced by histamine.

French studies reported in the September 25, 1986, issue of *La Presse Médicale* show that ginkgo has properties that alleviate allergies, but to date there has been no solid laboratory evidence to guarantee that *Ginkgo biloba* is effective against allergies in practice. However, there is good reason to include ginkgo in the long-term management of allergies because we know how ginkgo works: It is a free-radical scavenger that reduces inflammation. I recommend ginkgo to patients who have allergies, ranging from mild cases to severe and chronic ones. Allergy sufferers should give ginkgo a chance by taking 120 to 240 milligrams daily for approximately three months to see if their allergic conditions improve.

GINKGO AND ASTHMA

Asthma has plagued humankind at least since the beginning of civilization. It was known to the ancient Egyptians. The word itself comes from the Greek word *asthma,* meaning "panting."

Asthma is a respiratory disease in which the bronchial passages, the small air passages to the lungs, become inflamed and make breathing difficult. Often, asthma starts in childhood and clears up with time, but asthma can begin at any age. Unfortunately, it is on the rise. The number of cases, and asthma-related deaths, increased by 40 percent in the thirteen years between 1982 and 1995, for reasons that are still unclear.

There are two types of asthma: extrinsic and intrinsic. Extrinsic asthma is triggered by an allergy to something in the air, such as pollen or dust. Intrinsic asthma usually develops in adulthood, and may start with a respiratory infection or, sometimes, an emotional trauma, such as the loss of a loved one.

During an asthma attack, the bronchial passages narrow, and the mucous lining swells. These reactions combine to restrict the flow of air to the lungs and make breathing difficult. Attacks can be as mild as a dry cough or as severe as wheezing, with rapid heartbeat and an inability to speak. Some attacks can be fatal.

At present, there is no cure for asthma, but Western doctors try to control it with various oral or inhaled drugs. Severe asthma is treated with corticosteroid drugs, often taken through an inhaler. Over the long term these medications can have punishing side effects, including swelling, high blood pressure, glaucoma, and suppression of the body's natural hormones.

For centuries, the Chinese have treated asthma and bronchitis with ginkgo. The ginkgo seed has an astringent property, an ability to dry out wet, runny conditions, such as asthma and chronic diarrhea. Chinese herbalists put ginkgo seed in soups and other foods, and sometimes brew it as a tea.

There is some evidence that modern medicine should take a serious look at the healing properties of *Ginkgo biloba* as a treatment for asthma. First of all, if the asthma sufferer is taking steroids, the herbal extract may allow for lower doses of steroids, thereby lessening some of the side effects that steroids produce. Futher, steroids cause people to bruise easily because these drugs affect the walls of the blood vessels and make them fragile. Flavonoids such as those contained in ginkgo protect blood vessels, strengthening them so that they are less likely to break and cause bruising.

Ginkgo biloba may diminish a tendency to develop asthma through its action in fighting platelet-activating factor (PAF). This is probably the work of the ginkgolides in ginkgo (see Chapter 3). PAF causes platelets to stick together and form clots. It worsens allergic inflammation and may promote bronchial constriction, decreasing the intake of oxygen and causing the breathing problem that asthmatics know so well.

Over and above all, we know that *Ginkgo biloba* is a free-radical scavenger that reduces inflammation. I recommend it to people with chronic asthma, whether it is mild or severe. They should take 120 to 240 milligrams daily for approximately three months and see whether there is improvement. A doctor should monitor the condition, particularly if asthma is acute. Of course, as people experience the results themselves, they should keep notes on the condition day by day to see not only what medications and dosages are effective, but how other factors—weather, the season of the year, and the pollen count—affect the condition. It is good, too, to keep abreast of new research being done on *Ginkgo biloba* in the area of asthma and bronchial conditions.

GINKGO AND HEPATITIS

There are several types of hepatitis, each with various causes, including heavy drinking, drug abuse, and, more commonly, viruses that attack the liver. The different types of hepatitis are type A, type B, type C, type E, and the delta hepatitis virus, and it is likely that more will be discovered.

These types are found in varying degrees in different areas of the world and are transmitted differently.

Type A is thought to be spread through an infected person's excrement. The virus contaminates food or drinking water as a result of poor sanitation facilities or practices, and is passed from an infected individual to other people in this manner. According to the American Medical Association's *Encyclopedia of Medicine*, most people have already been infected with hepatitis A, often without knowing it, and have developed a resistance, or immunity, to further attacks. Good hygiene, especially around food, is the best preventive against infection by hepatitis A.

Many people in the developing countries—an estimated 20 percent, according to the World Health Organization—are carriers of the hepatitis B virus. Many who are carriers develop the disease, which causes the liver to produce fibroids—benign tumors that consist of fibrous and muscular tissue—so that the liver cannot function properly. Without a functioning liver, a person cannot live. The liver is the major chemical plant of our body. It detoxifies a lot of toxins, whether ingested or produced by the body in the course of normal metabolism. The liver also metabolizes fats, stores vitamins and minerals, activates enzymes, and produces proteins for the blood, among other things.

Type B hepatitis is spread from one person to another chiefly through blood transfusions and sexual contact (although tests and screenings have minimized the risk of being infected through transfusions). This type is more serious than type A and leads to chronic problems, which can include liver cancer and cirrhosis. In the United States, the rate of infection for type B is less than one percent. However, travellers to developing nations are at risk, as are homosexuals, intravenous drug abusers, and children born to mothers who are carriers. A vaccine has been developed for type B hepatitis. Vaccination is recommended for those at high risk of developing the disease, as well as for all newborns.

Chinese researchers studied the effect of *Ginkgo biloba* on the tissues of the liver. They looked at eighty-six people who had chronic hepatitis B with early fibrosis development. They measured a number of enzymes that are generally associated with development of fibrosis. They took biopsies of the subjects' livers and studied them with light-transmitting microscopes and electron microscopes. After three months of treatment with *Ginkgo biloba,* levels of four major enzymes associated with fibrosis were significantly reduced. The patients, all from the developing countries, had a remission of fibrosis from chronic hepatitis B, according to the report on the study published in *Chung-Kuo Chung Hsi i Chieh Ho Tsa Chi,* October 15, 1995.

The fact that researchers have found that *Ginkgo biloba* has a positive effect on hepatitis B is not surprising. Hepatitis is an inflammation of the liver that damages that organ's cells. We have seen over and over again that *Ginkgo biloba* is an effective fighter of inflammation and that it promotes the life of cells. This encouraging result is preliminary, but the benefits of *Ginkgo biloba* in treating hepatitis clearly merit further study. Meanwhile, it offers hope to those who fear they are at risk for the disease. Those travelling to developing nations should prepare for their trips by taking extra doses of *Ginkgo biloba* extract.

SUMMARY

Because it fights the breakdown of the cells, *Ginkgo biloba* slows the aging process and the even more damaging results of exposure to ultraviolet light from the sun. In treating those exposed to the nuclear disaster at Chernobyl in the former Soviet Union, scientists discovered how truly potent *Ginkgo biloba* is in fighting free radicals and the breakdown of the cells due to radiation exposure. Although it is to be hoped that no one will ever again be subjected to such intense radiation, everyone on the planet is exposed to ever-mounting levels of radiation. *Ginkgo biloba* is able to counteract the slow, cumulative effect of radiation we absorb through our food supply and from x-rays and other developments of modern life.

As I have noted repeatedly, *Ginkgo biloba* is one of the most effective scavengers of free radicals, those "incomplete" molecules or atoms that react to anything they come in contact with. They break up or bind with other molecules in the body, and are said to play a role in degenerative diseases like cancer. They are also the major cause of the aging process. Even age-old plagues like allergies may be offset by a consistent plan of taking *Ginkgo biloba*. Finally, we have seen that the herbal extract may well be useful to those who suffer from asthma and hepatitis.

In the next chapter, we will take up the various ways to incorporate *Ginkgo biloba* into a regular health plan, and how to evaluate the various products available on the market.

CHAPTER 9

A Guide for the Consumer

We have seen how *Ginkgo biloba* works and what it can do. Now let us turn to the following questions:

- Who should take *Ginkgo biloba* and why?
- When should they take it?
- How should they take it?

This chapter should help you to evaluate your own needs and come up with the answers that are best for you.

REASONS FOR TAKING GINKGO BILOBA

We know that the benefits of *Ginkgo biloba* seem to offset problems of aging—such as poor memory and leg cramps. Does that mean that young people should not bother to take it? If you are in good health, with normal blood pressure, do you need to take it? What is the difference between using *Ginkgo biloba* as a therapy and as a preventive?

There are specific times when you would have reason to take ginkgo, such as when you have been exposed to too much sun, or when you need to be alert to study for exams, or if you fear the effects of exposure to radiation. There are also certain categories of people for whom regular use of ginkgo might be recommended. Anyone who worries about any of the following conditions should take ginkgo.

Mental Alertness

People who want to keep their minds sharp, particularly in stressful environments, could probably benefit from *Ginkgo biloba*. Ginkgo keeps the brain healthy and the blood vessels that serve it strong and supple, and has been shown to improve short-term memory.

The Health of the Blood Vessels

Ginkgo biloba is a tonic to blood vessels. Research suggests that we benefit from it because it prevents or repairs microscopic damage to the cells and vessels before they can accumulate into large problems.

Cardiovascular Risk Factors

Megadoses of ginkgo given in medical situations or experimental clinics have shown promising results. German scientists in 1992 administered high dosages (240-milligram tablets daily for twelve weeks) to twenty patients with various diseases that led to dangerous blood clotting. These patients, who were not hospitalized, had coronary heart disease, hypertension, high cholesterol, and diabetes. The clotting factors decreased in every one of the patients taking *Ginkgo biloba*, leading the researchers to report: "The medication can thus positively influence these cardiovascular risk factors over the long term."

The Health of the Skin

Data indicate that *Ginkgo biloba* keeps the skin not just healthy but youthful. The skin ages when it is starved of nutrients and oxygen, which happens when the blood vessels are constricted. Thermography would provide a quick demonstration of this. This technique records the temperature patterns of the skin. Each different temperature registers as a different color, with a reddish-orange warm zone being the best. As a thermogram shows, if you puff a cigarette, the temperature of your extremities, such as the fingers, drops a few degrees because the arteries are constricted. That also means that for a period of time lasting a few minutes, the supply of nutrients and glucose to the tissues of the fingers is reduced.

It is understandable that if you don't feed an organ, it will suffer and eventually wither, and the skin is an organ. *Ginkgo biloba* helps skin cells stay strong and beautiful.

Cerebral Insufficiency

In Germany, as we noted, *Ginkgo biloba* extract is licensed for the treatment of cerebral insufficiency, which encompasses a constellation of problems including impaired memory, dizziness, ringing in the ears, headaches, nervousness and anxiety. Ginkgo has been shown to bring about improvement in such symptoms in many cases.

Problems with Balance or Equilibrium

Problems with balance or equilibrium occur frequently, and most of the time they cannot be attributed to a single cause. Sometimes doctors administer extensive, expensive tests and come up with no cause. But if you start treatment with *Ginkgo biloba* early and take it for three months, you may experience improvement. In fact, experts found improvement in 80 percent of such cases.

The Effects of Radiation

Clastogenic factors found in the blood of people exposed to radiation break chromosomes and severely damage cells (see Chapter 3). The greater the dose of radiation received, the higher the number of clastogenic factors in a person's blood. *Ginkgo biloba* is an antioxidant that scavenges damaged molecules. It has been shown to decrease—or even eliminate—

clastogenic activity of the plasma in people who were exposed to high levels of radiation. People who have been, or who know they will be, exposed to radiation, such as those undergoing radiation therapy, can protect themselves from the bad side effects, without diminishing the good effects, by taking *Ginkgo biloba* extract.

Failing Eyesight

Without proper flow of blood to the retina, the layer of nerve tissue lining the inside of the back of the eye, one loses vision and eventually becomes blind. Research has shown that *Ginkgo biloba* can bring about an improvement in the health of the retina, benefitting damaged areas without affecting healthy tissues. Further, damage to the visual field caused by a chronic lack of blood flow may be reversed to a significant extent by taking *Ginkgo biloba*.

METHODS OF TAKING GINKGO BILOBA

Ginkgo is available in dried leaf form; as a tincture, in homeopathic preparations; and in extracts. Obtaining an extract of the total herb is what is important. The whole herb is active while single isolated components are not as effective. This is not always the case with all herb-derived medicines. The potent heart medicine digitalis, for example, is an ingredient found in the *mao ti huang* plant, which English-speaking people call foxglove. In the case of digitalis, as in the case of many other drugs, the active ingredient of the plant is more important than the entire herb. However, in the case of herbs, like *Ginkgo biloba*, extracts of the whole leaf are more effective than extracts of glycosides and terpene lactones which are the active components it contains.

Traditional Chinese Methods

Chinese herbalists use three parts of the *Ginkgo biloba* tree—the nuts (sometimes called the fruit), the seeds, and the leaves. The gingko leaf has different medicinal uses from those of the seed. It is an antioxidant with vitamins and minerals. When the leaves turn yellow, Chinese herbalists soak them in rice wine for three weeks and then take a few tablespoons at a time to improve circulation and mental function.

The nuts contain terpenes that irritate the skin. Contact with ginkgo fruit, a plumlike fruit that matures in late autumn, can cause skin rashes and itching. Swallowing even a small piece of the fruit or its seeds results in painful spasms. The Chinese boil the fruit to remove these properties, and they add the nuts and the seeds to soups as a medication for "drying" wet, runny conditions such as asthma and chronic diarrhea.

The Chinese make ginkgo tea to treat asthma and bronchitis. This has worked for centuries, although results have varied according to the quality of the leaf and growing conditions. Droughts, heavy rains, heat spells, and a variety of other natural phenomena affect the quality and the quantity of the active ingredients in the plant.

The Chinese use a ginkgo paste on wounds. They have seen for years that it prevents infection. We now know that it kills germs because it contains ginkolic acid. I doubt that there is anything seriously wrong with applying ginkgo externally, as traditional Asian herbalists do for some conditions, but since *Ginkgo biloba* contains a substance that can create contact dermatitis, I do not think it is a good idea. It is estimated that 30 percent of the population would develop rashes from ginkgo applied to the skin, so I would recommend getting the benefits of ginkgo by taking it orally in tablets of *Ginkgo biloba* extract.

As an Extract

Modern producers standardize extracts from ginkgo leaves and turn them into tablets, liquids, or intravenous preparations. Extracts are among the most commonly prescribed drugs in France and Germany.

Dr. Willmar Schwabe of Schwabe GmbH, the largest phytomedicine company in Germany, began to research ginkgo more than two decades ago. He developed an extract of ginkgo leaves known as EGb 761. The process of making the extract begins in the age-old Chinese manner: The leaves are harvested in the late summer, when their active compounds are at their greatest levels and thus most beneficial. The modern, scientific process of isolating the active ingredients and standardizing the extract so that its strength will be consistent takes two weeks and involves twenty-seven steps. Some fifty pounds of leaves yield one pound of the extract, called *Ginkgo biloba* extract, or GBE, which in standardized form contains 22 to 27 percent ginkgo flavone glycosides, or flavonoids, and 5 to 7 percent terpene lactones. Standardized *Ginkgo biloba* extract is also available in drop form, with a content of 9.6 milligrams of flavone glycosides and 2.4 milligrams of terpene lactones per dose.

Companies preparing ginkgo extract check the quality of the leaves on a regular basis and harvest them when the concentration of active ingredients in the leaves is optimal. Then they dry the leaves until three-fourths of their water has evaporated. The leaves are then pressed firmly to prevent fermentation and absorption of water from the environment. Then they go through a microwave tunnel. There are no organic solvents used in the production of *Ginkgo biloba* extract.

Several factors affect the potency of the ginkgo plant, including where ginkgo leaves are grown, the conditions under which they are harvested, and the harvesting method used. The quality of the active components can vary by as much as 300 percent, depending on the location, time of year, and harvesting method. By taking something that has been around for thousands of years and standardizing it so that it has a constant strength and composition, we have helped humankind. The modern extraction and standardization process is an improvement over brewing something in the kitchen, a process that is subject to variation and inconsistencies—you may pick the wrong leaves at the wrong time; the leaves may be affected by drought; the concentration may vary from batch to batch. Standardized extract makes a good thing more consistent.

Guidelines for Potency

The dose of *Ginkgo biloba* extract tablets used in most clinical studies is 40 milligrams of standardized extract three times per day over a period of at least eight weeks. The standard intravenous dose is 0.5 milligrams three times a day. Taking high doses for indefinite periods does not seem to be required even for serious conditions like damage caused by radiation exposure. While some serious conditions are treated with high doses, most studies indicate that a daily dosage of 120 to 240 milligrams of standardized *Ginkgo biloba* extract daily is effective over the long term, with improvement in various conditions appearing after eight to twelve weeks. Most preparations on the market are tablets or capsules of smaller dosages, so they must be taken three times a day to equal the recommended daily dosage. However, there are products available that need be taken only twice, once in the morning and once at night, so that you do not need to worry about it during the work day.

Researchers treating or studying "cerebral insufficiency," in its various forms from memory loss to depression, treat patients with 120 milligrams of *Ginkgo biloba* three times a day in tablet or drop form. In treating or experimenting on leg cramps, vertigo, and ringing in the ears, a higher dose—160 milligrams daily, divided into two or three equal doses—of the standardized extract is used.

Sometimes results may take up to twenty-four weeks (six months) to appear, which seems to indicate that areas that have long been deprived of oxygen and nutrients revitalize more slowly. Take the case of the three-month French study on patients with vertigo (see Chapter 6). Balance was restored slowly; it was noticeable after thirty days and significant after sixty days. The big difference, however, was seen after ninety days. In the case of the workers who cleaned up the Chernobyl nuclear plant (see Chapter 8), the beneficial effects of ginkgo lasted for seven months, even in people who had not taken any treatment beyond the initial ginkgo study period. The implication is that you need to take *Ginkgo biloba* only at certain intervals in the case of some chronic conditions, which can save you money. Above all, its long-lasting effects show how potent the antioxidant effect is.

The Question of Side Effects

Ginkgo extract is basically free of side effects. In more than forty-four research trials involving 9,772 subjects, many of which took place over a relatively long term, no serious problems were reported. There were some reports of mild gastrointestinal disorders, headaches, and allergic skin reactions, but these were rare. Fewer than one percent of those studied had mild stomach upsets or occasional headaches. (In those treated for problems involving circulation to the brain, the headaches may denote stimulation of damaged vessels.) There were no drug interactions, even though ginkgo was used in patients who were often taking other drugs. No serious negative reactions have been reported in patients taking as much as 600 milligrams in one dose. Further, ginkgo appears to pose no danger to pregnant women or nursing mothers. There appears to be low risk in taking ginkgo over long periods of time.

A RECOMMENDATION

One particularly good form of *Ginkgo biloba* extract on the market is called BioGinkgo, manufactured by Pharmanex. This product has been formulated to make the most of ginkgolide B (see Chapter 3). It comes in two strengths, BioGinkgo 24/6 and BioGinkgo 27/7, an extra-strength version. BioGinkgo 27/7 provides the highest amount of flavonoid glycosides and terpene lactones mandated in standardized extract: 27 percent and 7 percent, respectively. A scientific study published in *Planta Medica* in 1997 showed that it enters the bloodstream more quickly than other extracts and remains active for a longer time.

Unlike other ginkgo extracts, BioGinkgo is taken only twice a day—morning and evening—with food or liquid.

The tablet is coated to make it easy to swallow. Some will see benefits in two or three weeks, but what researchers like to call "optimum benefits" will appear in twelve weeks.

SUMMARY

It is an unfortunate but unavoidable fact of life that we are all going to age and, ultimately, we are all going to die. However, the choices we make can have a significant impact on both the length and the quality of our lives. My personal belief is that, if you want to live longer and have a more active life, you should start taking ginkgo at the end of puberty, some time between the ages of seventeen and twenty-five. There are centuries of evidence to support doing so. There is no contraindication for doing so, no reason why we should not do so.

People in Asia have been consuming ginkgo as part of their food supply from the time they are old enough to eat solid foods. Ginkgo nuts are eaten in soups or as a delicacy almost every day. Many people also take all kinds of herbal remedies as part of their diet—but we do not live in parts of the world where this is common.

All of us are subject to toxins in the air we breathe. A polluted environment, whether self-made through smoking, or due to things humankind as a group has done to the earth, such as the release of radioactivity and petrochemicals, is a good reason to take *Ginkgo biloba*.

You, and your doctor if you are under treatment, might test the following system. Take 120 milligrams to 240 milligrams of *Ginkgo biloba* extract daily for three months. Take notes about how you feel and any unusual occurrences. Together, you can decide whether ginkgo is beneficial for you.

In the next chapter, we will conclude our findings, not just about *Ginkgo biloba*, but about exciting new developments in complementary medicine.

CHAPTER 10

Conclusions

In reading this book, you have taken an important step in improving your health, your well-being, and your prospects for a happy, vital life that will extend well into your old age. *Ginkgo biloba* extract offers the greatest hope for the regeneration of our bodies since Western science discovered the benefits of ginseng and vitamin E. Like these dietary supplements, *Ginkgo biloba* extract should be, in my opinion, a regular part of anyone's health program.

In this book I have discussed some of the conditions that *Ginkgo biloba* is said to improve. Just paging through the earlier portions of the book will show you that these ailments range from poor memory to clogged arteries to ringing in the ears. Over the course of the months I spent writing this book, at least three events relating to the health problems we have looked at here made the news.

First, there were reports from Long Island, New York, that radioactive leakage at a research facility was slowly seeping into the ground around the facility. Then there were reports of two fires at a nuclear power plant seventy miles northwest of Tokyo. (In both cases, officials reported that the radioactivity released was well within "safe" levels for those who came in contact with it.) Finally, a noted actress had to withdraw as the hostess of an awards show because of diabetic retinopathy—her vision was temporarily blurred due to complications from her diabetes. Her spokesperson said, "There are a number of side effects to diabetes, and this is one of them. Small blood vessels in the back of your eye can burst or leak blood into your eye. It can be serious."

All of the health problems associated with these news stories could have benefitted from *Ginkgo biloba*.

A BACKWARD LOOK

In this book we have covered millions of years, because that is how far *Ginkgo biloba* goes back—all the way to the Triassic period. This makes it what Charles Darwin called a living fossil, a plant that grows today much as it did when its seedlings were trod underfoot by the dinosaurs. Ginkgo had the last laugh there, because it survived those fabled beasts.

The ginkgo tree thrives in inhospitable conditions, including drought, smog, frost, and low sunlight. This prehistoric plant reigns in our most congested urban environments and may counteract pollution. The proof of the resilience of the ginkgo tree is that it is the most widely planted tree in New York City.

Ginkgo biloba has come to the attention of the Western world only recently. Europeans have embraced it for a variety of physical problems, particularly for brain-related conditions, and their medical doctors prescribe it. The United States has been a little slower to catch on, which is one of the reasons I wrote this book.

Western medical scientists have begun conducting research on ginkgo, but Asian herbalists report fifty centuries of successful results with it.

WHAT GINKGO DOES

The specific benefits of *Ginkgo biloba* all derive from its three essential actions in the body:

- Cleaning up free radicals, those molecules or fragments of molecules that react with and destabilize other molecules, causing cells to break down and age.

- Keeping nutrients flowing to all cells of the body by keeping the blood vessels strong and supple.
- Fighting the damage ultraviolet rays cause our bodies, including the cells of the skin and the retina of the eye.

Of course, there is no guarantee that what works even for your twin will work for you. Everyone is different, but the data on people from different ethnic and economic backgrounds and conditions of life and diet consistently indicate that *Ginkgo biloba* is good for you. Ginkgo may not completely cure something that has been damaged, but there is a possibility of minimizing the condition.

Ginkgo biloba helps to prevent the aging of the skin, and the loss of hearing and vision. It helps people exposed to radiation, including sunshine, especially at high altitude, which destroys our DNA and promotes the development of cancer. It prevents the cells from changing. We have seen that it combats depression that is organically caused. In Germany, *Ginkgo biloba* is approved as a supplemental treatment for hearing loss.

Over time, ginkgo extract has even more enduring effects on the heart, vision, skin, and lungs, because flavonoids such as those it contains seem to be most effective on organs with a lot of connective tissues, like the aorta, eyes, skin and lungs. The effect of *Ginkgo biloba* on the cells of body tissues is profound. It stabilizes cell membranes, the covering that holds the components of the cells together and keeps out substances foreign to the cells. *Ginkgo biloba* is one of the most effective scavengers of free radicals, those molecules or atoms that react with almost anything they come into contact with.

Ginkgo biloba offers help for problems with the reproductive system. Because it is so effective in treating blood vessels, it offers hope to men who suffer from impotence due to the degeneration of the blood vessels of the penis. French scientists discovered that *Ginkgo biloba* effectively relieves extreme symptoms of premenstrual syndrome related to water retention.

Western science has managed to prove what the Chinese have believed for centuries—that *Ginkgo biloba* helps mental performance and improves inflammatory conditions, such as asthma. When any kind of mental function—whether it is memory, alertness, attention, disorientation, or any other mind state—is slowed because of poor circulation to the brain, or even when a person is depressed, *Ginkgo biloba* extract provides an effective way of improving these conditions.

The distracted, hurried, sometimes cutthroat pace of modern life makes people intolerant of those who are slower than they. People may show compassion for individuals who are obviously handicapped, but not for those who do not think as quickly as others, who do not perform well intellectually. Try doing less than average work a few days at the office and see how your colleagues treat you! Yet it is part of the evolutionary process for the body to break down with age. From an evolutionary point of view, when you reach the age when you can no longer reproduce, you are no longer important to the survival of the species, so your organism begins to shut down.

Aging affects every cell of the body. Our cells divide and reproduce throughout life, but with age these divisions slow down. We see this with the skin as we age. In addition, free radicals attack and damage the cells, and the actions of our immune systems can turn against our own tissues as well. All of these factors accelerate the aging process.

As a society, we have little tolerance of those who are getting older. The fact that recent rounds of layoffs (conveniently softened under the new word *downsizing*) hit particularly hard at those over forty shows us this. Particularly in the workplace, we need to prevent the degradation of our mental functions. We need to be alert for signs that our memory is slowing down, and we need to treat such problems once they occur. In part, that entails proper diet and sleep, and cutting back on the consumption of tobacco and alcohol, but it also involves such supplements as *Ginkgo biloba* extract.

Symptoms related to aging are very individual, and the same symptoms can result from different causes in different people. Ginkgo addresses many of these different conditions, whether we know what is causing them or not.

Many older people suffer from problems with memory, mood, and concentration, and from headaches, tinnitus, and loss of equilibrium. Sometimes all this occurs at the same time. Sometimes it occurs slowly, or just for a short time. An insufficient supply of blood to the brain is responsible for an estimated 20 percent of such cases, and for 50 percent in cases of degeneration from Alzheimer's disease. Ginkgo can help, at least to a certain extent, in the early stages of Alzheimer's disease.

WHY GINKGO WORKS

Ginkgo biloba extract has been proved to help memory, improve general brain function, and counteract problems with blood and circulation. *Ginkgo biloba* has various important medical uses because it benefits the blood vessels, and for that reason benefits the performance of various organs, most notably the brain. Its many healing properties are also attributed to the complex structure of its molecules and to the flavone glycosides and terpene lactones it contains.

The flavone glycosides are antioxidants. *Ginkgo biloba* extract has been proved to be more effective as an antioxidant than beta-carotene (which can be converted by the body to vitamin A) and vitamin E. The flavone glycosides also protect the cells against the breakdown of arachidonic acid, which keeps the membranes of the cells healthy and permeable. The terpene lactones confer another type of necessary help to the body. They offset the actions of platelet-activating factor (PAF), a substance that causes the platelets of the blood to clump together and promotes inflammation in tissues. Although the various active elements of *Ginkgo biloba* each play a separate role, experimental studies show that the action of all these elements together makes them more effective than any of them would be alone. Further, these compounds are found only in the ginkgo tree. Their chemistry and molecular structure are absolutely unique. The medical benefits of *Ginkgo biloba* supplements depend on the proper balance of the two active components—ginkgo flavone glycosides (also called flavonoids), at levels of 22 to 27 percent, and terpene lactones, at levels of 5 to 7 percent.

GINKGO VERSUS DRUGS

In some cases, *Ginkgo biloba* may not be more effective than currently available drugs, but because it has no serious side effects and shows no problems with interactions with other medications a patient may be taking, it is probably superior to many medications and vitamins. In other cases, *Ginkgo biloba* is more effective than many drugs, since it is effective against all of the following:

- Damage to blood vessels.
- Slowing of the metabolism.
- Thickening of the blood.
- Decreased brain activity.

Ginkgo improves the irrigation of tissues that have been deprived of oxygen. It keeps the arteries functioning. It is an anti-clogging agent.

TAKING PERSONAL RESPONSIBILITY FOR HEALTH

Educated Americans are not waiting to get the green light from their medical establishment. Studies show that better-educated people are taking responsibility for their own health. They do not see their doctors as authority figures, but as partners in their health. These are not people who are susceptible to charlatans and snake-oil salesmen. Significantly, the more affluent, who are the best able to pay for expensive health care, are the most likely to be interested in new methods, most of which are more easily affordable. A case in point is Chinese acupuncture, which came to the attention of the West in the early 1970s. A few treatments, which then cost as little as twenty-five dollars apiece, were often more effective than expensive long-term drug regimens for arthritis and colitis—and there were no side effects.

Many people with persistent asthma, arthritis, cancer, chronic pain, AIDS, and other debilitating diseases are in the vanguard of this attitude toward health care. They want to assume a role in their own care in order to have more control over their own fate.

Others who currently enjoy good health know that the best way to fight disease is to prevent it. One good example is dysentery. We essentially eradicated this disease not because of superior drugs, but because we learned the value of clean water, and of clean wells and pipes to carry clean water.

The hallmark of the twenty-first century could well be the rational use of herbs, rather than overuse of drugs, to restore and promote health. Drugs can cure disease, but herbs can promote health. No reader should misunderstand the fact that the biomedical system, in which the United States leads, has made countless advances, among them polio and other vaccines. In the last quarter-century, quantum leaps have been made in the care of heart patients. The problem is that the patients with the most money, those who can reach the best health-care facilities, are the best able to take advantage of these advances. The key to improved health for all is to make the most of the knowledge of both the ancient East and the technological West.

If we truly wish to reduce our health-care costs and to live vital long lives on our own—without being a burden to others or losing our independence—we must seize control of our well-being. Prevention, not only cure, must be our goal.

BRINGING HEALTH TO THE MARKETPLACE

One of the challenges to improved health care is the amount of time and money required to get a medication approved in the United States. Paradoxically, this system was set up to protect the public, and at times it has, such as in the famous case of thalidomide, which caused birth defects in Europe and Canada in the early 1960s. However, our federal approval process ignores or dismisses prior use—that is, evidence gathered over the years in other nations and in other systems of medicine. *Ginkgo biloba*, for example, which has proven itself to be effective over a period of nearly 5,000 years, is not yet certified by the U.S. Food and Drug Administration.

Europeans, in contrast, accept the evidence of centuries. *Ginkgo biloba* is marketed over the counter in lower dosages on that continent, and prescribed in higher doses by physicians. Europeans, particularly the French and Germans, acknowledge its proven history as a boon to humankind, and scientists in Europe have studied it in labs. In this country, *Ginkgo biloba* extract is readily available in health-food stores, drugstores, supermarkets, and other stores, but no specific disease-related claims can be made for it. Consumers have to hear or read about it and its benefits in order to try it for themselves.

Our advanced biomedicine used to frown on herbs and alternative medical practices like chiropractic and nutritional supplementation, but now it is looking at these areas more objectively. In fact, the National Institutes of Health now calls these approaches *complementary medicine*. Their ancient wisdom, based on centuries of trial and error, is finally being considered seriously. And why not? Somehow we forgot that "complementary" medicine was the basis for everything modern medicine has achieved. Willow bark was used before we knew about salicylates and the action of aspirin on prostagladins. Belladonna and poppy were taken before acetylcholine and endorphins were understood. People used these herbs because they observed that these plants worked.

The benefits of herbal medicine, particularly *Ginkgo biloba*, need to be established in this country. *Alternative Medicine: Expanding Medical Horizons*, the 1992 report on alternative medicine presented to the National Institutes of Medicine (see Chapter 2), said as much. It recommended that the connection between a patient's mind and body be acknowledged and treated with greater respect in the healing process. It further said that herbs, vitamins, and manual healing methods (chiropractic) should be studied in a systematic way. American medical culture has come a long way since it scoffed at vitamins and the idea that diet plays a role in the onset of disease.

FINAL CONCLUSIONS

It is clear to me that *Ginkgo biloba* extract should be as much a part of a person's health regime as vitamin supplements. In fact, the bioflavonoids in *Ginkgo biloba* even improve the absorption of vitamin C. The nutritionist Adelle Davis said in her book *Let's Eat Right to Keep Fit* (Harcourt Brace Jovanovich, 1970), "Experiments indicate that . . . bioflavonoids reduce the need for vitamin C and make it more effective, thereby increasing the strength of capillary walls, reducing inflammation, and decreasing the seepage of blood cells and proteins into the tissues."

All evidence points to the conclusion that *Ginkgo biloba* extract will one day be as well accepted as vitamin E, but I hope readers of this book won't wait for that. I hope you will consider taking *Ginkgo biloba* now as a preventive medicine to fight the breakdown of skin and retina cells caused by ultraviolet rays, and to boost your mental performance. No significant side effects have ever been reported, even in megadoses, so take it for three to six months and keep a little notebook to record how you feel. Particularly at the end of thirty-six weeks, I think you will see a difference, and, in years to come, I think you will be happy that you learned about *Ginkgo biloba* extract.

Glossary

antioxidant. A substance that neutralizes free radicals that break down the molecules and cells of the body.

arachidonic acid. An essential fatty acid that is a constituent of human cell membranes. It is also a precursor of inflammatory body chemicals.

arterial thrombosis. A condition in which a blood clot in an artery obstructs blood flow.

arteriosclerosis. Hardening of the arteries.

arteritis. Inflammation of an artery wall that reduces blood flow and sometimes promotes blood clots.

atherosclerosis. A thickening and narrowing of an artery caused by the deposit of fatty plaques on the interior walls. This condition causes more deaths in the United States than any other condition.

Ayurveda. The ancient medical system of India.

cardiovascular system. The heart and blood vessels.

central nervous system. The brain and spinal cord.

claudication. Cramping pain in the legs, which can lead to a limp; its usual cause is spasm of the arteries.

CNS. Central nervous system.

cochlea. An organ in the inner ear that transforms sound vibrations into nerve impulses.

cytotoxic. Damaging or toxic to cells.

dementia. General loss of mental function, which results in memory loss, confusion, emotional outbursts, and, often, embarrassing behavior, such as incontinence.

diabetes. A condition in which the pancreas produces too little or no insulin, the hormone that enables the cells to absorb glucose. The body becomes unable to use glucose, resulting in major diverse damages, mostly in blood vessels and sensory organs.

drug. According to the definition of the American Medical Association, "a chemical substance that alters the function of one or more body organs or changes the process of a disease." Drugs include prescribed medicines, over-the-counter remedies, and illicit drugs such as cocaine.

edema. Accumulation of fluid in the tissues, which may or may not be visible and can be caused by injury, physical disorder, or leakage of fluids from capillaries or veins.

embolism. Blockage of an artery by a cluster of material circulating in the bloodstream. The particle, called an embolus, can be a blood clot, an air bubble, bacteria, cholesterol, or any of a number of other substances.

fibrinolysis. The breakdown of fibrin, the principal component of any blood clot.

free radical. A molecule or atom that readily reacts with any other molecule it encounters, changing the molecule in harmful ways and causing it to break down. Free radicals are believed to play a role in inflammation, in degenerative diseases such as cancer, and in the aging process.

glucose. A simple sugar that is the body's chief source of energy. Most of it is derived from carbohydrates in the diet, although interactions of fats and proteins in the cells produce a small amount.

infarct. Death of an area of tissue due to a lack of blood supply.

ischemia. Insufficient supply of oxygen to tissues in the body. Used most often of heart muscle or brain tissue.

lipid. Any of a group of fatty substances including triglycerides (the main forms of fat in body fat), phospholipids (vital constituents of cell membranes), and sterols such as cholesterol.

mitochondria. Small, sausage-shaped structures in the cells that are responsible for producing energy.

myocardium. The middle muscular layer of the heart wall.

PAF. Platelet-activating factor.

phytomedicine. A plant-based medicine.

placebo. A chemically inactive substance given instead of a drug. In clinical trials, placebos result in clinical improvement in up to 60 percent of patients.

peripheral vascular disease. Constriction of the blood vessels in the legs and in the arms, usually causing pain and cramps; it usually starts with inflammation of the walls of arteries.

plasma. The yellowish fluid portion of the blood, minus the red and white blood cells. It consists of water, salt, nutrients, and proteins.

platelet-activating factor. A substance that combines with the surface of platelets in the blood and causes them to form clots. It is also associated with inflammation in the skin, and possibly in the bronchi.

prostaglandin. Any of a group of naturally occurring body chemicals, derived from fatty acids, that have a variety of effects, including contraction, inflammation, and damage to tissues.

Raynaud's disease. A condition in which exposure to cold temperatures causes the blood vessels of the fingers and toes to constrict, cutting off the blood supply.

retina. A light-sensitive layer of nerve cells that lines the back of the eye and converts light to nerve impulses, resulting in sight.

tinnitus. A condition characterized by ringing, roaring, hissing, and/or buzzing noises in the ears.

About the Author

Georges M. Halpern was born a French citizen in Warsaw, Poland, and attended medical school at the University of Paris. In 1964, he received his M.D. degree and was awarded a silver medal for his thesis on the histamine-releasing properties of colistin. He subsequently qualified in nuclear medicine and was board certified in internal medicine and allergy. In 1992, he received his Ph.D. degree, with highest honors and jury honors, from the Faculty of Pharmacy at the University of Paris XI—Chatenay Malabry, for his thesis on skin hyperreactivity and histamine.

Until 1997, Dr. Halpern was an adjunct professor of medicine at the University of California–Davis, Division of Rheumatology, Allergy, and Immunology, and a professor of nutrition, College of Agricultural and Environmental Sciences. He is currently professor emeritus.

Dr. Halpern has published eleven books, fifty-nine book chapters, 195 original papers, and hundreds of reviews and abstracts. He has lectured in seventy-two countries. In 1985, he was awarded the Medal of Vermeil by the city of Paris for his "outstanding contributions to medicine, dedication to patients, and personal achievements."

Dr. Halpern is a Fellow of the American Academy of Allergy, Asthma, and Immunology, and of twenty-six other academies and scientific societies. He is also an officer in the French National Order of the Merite Agricole for his original contributions to French cuisine.

Georges M. Halpern is married to Emiko Oguiss, and has two daughters, Emmanuelle and Emilie.

Index